Geometry and Topology in Music

This book introduces path-breaking applications of concepts from mathematical topology to music-theory topics including harmony, chord progressions, rhythm, and music classification. Contributions address topics of voice leading, *Tonnetze* (maps of notes and chords), and automatic music classification.

Focusing on some geometrical and topological aspects of the representation and formalization of musical structures and processes, this book covers topological features of voice-leading geometries in the most recent advances in this mathematical approach to representing how chords are connected through the motion of voices, leading to analytically useful simplified models of high-dimensional spaces. It generalizes the idea of a *Tonnetz*, a geometrical map of tones or chords, and shows how topological aspects of these maps can correspond to many concepts from music theory. The resulting framework embeds the chord maps of neo-Riemannian theory in continuous spaces that relate chords of different sizes and includes extensions of this approach to rhythm theory. It further introduces an application of topology to automatic music classification, drawing upon both static topological representations and time-series evolution, showing how static and dynamic features of music interact as features of musical style.

This volume will be a key resource for academics, researchers, and advanced students of music, music analyses, music composition, mathematical music theory, computational musicology, and music informatics. It was originally published as a special issue of the *Journal of Mathematics and Music*.

Moreno Andreatta, Director of Research at CNRS/IRMA and Associate Researcher at IRCAM, France.

Emmanuel Amiot, Associate Researcher at UPVD, France.

Jason Yust, Associate Professor at the Boston University School of Music, USA.

Geometry and Topology in Music

Edited by
Moreno Andreatta, Emmanuel Amiot
and Jason Yust

CRC Press
Taylor & Francis Group
Boca Raton London New York

CRC Press is an imprint of the
Taylor & Francis Group, an **informa** business
AN AUERBACH BOOK

First edition published 2025
by CRC Press
4 Park Square, Milton Park, Abingdon, Oxon, OX14 4RN

and by CRC Press
2385 NW Executive Center Drive, Suite 320, Boca Raton FL 33431

British Library Cataloguing-in-Publication Data
A catalogue record for this book is available from the British Library

ISBN: 978-1-032-84515-9 (hbk)
ISBN: 978-1-032-84518-0 (pbk)
ISBN: 978-1-003-51361-2 (ebk)

DOI: 10.1201/9781003513612

Typeset in Times
by codeMantra

Publisher's Note
The publisher accepts responsibility for any inconsistencies that may have arisen during the conversion of this book from journal articles to book chapters, namely the inclusion of journal terminology.

Disclaimer
Every effort has been made to contact copyright holders for their permission to reprint material in this book. The publishers would be grateful to hear from any copyright holder who is not here acknowledged and will undertake to rectify any errors or omissions in future editions of this book.

Contents

Citation Information

The following chapters in this book were originally published in the *Journal of Mathematics and Music*, volume 14, issue 2 (2020). When citing this material, please use the original page numbering for each article, as follows:

Introduction
Introduction
Moreno Andreatta, Emmanuel Amiot, Jason Yust
Journal of Mathematics and Music, volume 14, issue 2 (2020), pp. 107–113

Chapter 1
Why Topology?
Dmitri Tymoczko
Journal of Mathematics and Music, volume 14, issue 2 (2020), pp. 114–169

Chapter 2
Generalized Tonnetze *and* Zeitnetze, *and the Topology of Music Concepts*
Jason Yust
Journal of Mathematics and Music, volume 14, issue 2 (2020), pp. 170–203

Chapter 3
Homological Persistence in Time Series: An Application to Music Classification
Mattia G. Bergomi and Adriano Baraté
Journal of Mathematics and Music, volume 14, issue 2 (2020), pp. 204–221

Notes on Contributors

Emmanuel Amiot, LAMPS, Université de Perpignan, France.

Moreno Andreatta, CNRS-IRMA, Université de Strasbourg, France; CNRS-IRCAM, Sorbonne Université, Paris, France.

Adriano Baratè, Music and Computer Science Laboratory, University of Milan, Italy.

Mattia G. Bergomi, Independent Researcher and Scientific Advisor, Milan, Italy.

Dmitri Tymoczko, Music Department, Princeton University, USA.

Jason Yust, School of Music, Boston University, USA.

Introduction

Moreno Andreatta, Emmanuel Amiot, and Jason Yust

This special issue focuses on some *geometrical* and *topological* aspects of the *representation* and *formalisation* of musical *structures* and *processes*. There are six words in the previous sentence which have been emphasized since they could work as explanatory guidelines for the present issue. More generally, they seem to offer to the "working" musicologist and music-theorist some conceptual and operational principles for approaching contemporary "mathemusical" research from a more philosophical and epistemological perspective. Shedding some light to this double underlying dimension may constitute a fruitful exercise, so we hope, for the reader of the *Journal of Mathematics and Music* and of the present issue in particular. Let us start with *geometry*, one of the oldest fields of mathematics and whose history accompanied the discussions on the relations between music and mathematics since the time of Greek philosophy. Geometry has a central place in the Pythagorean tradition, and particularly in Plato's philosophy, which is well expressed by the famous quotation that was supposed to be engraved at the door of the Plato's Academy in Athens: "Let no one ignorant of geometry enter." In Plato's perspective, geometry is surely far from what it should be according to its etymological meaning, which reduces the scope of the discipline to an "earth measure." Plato himself found such a name totally ridiculous, as he says in the *Epinomis*, the short dialogue that serves as an appendix to Plato's *Laws*.[1] As it has been pointed out by several scholars and commentators of Plato, starting from the great and influential music theorist Aristoxenus of Tarentum, geometry is first of all a science of abstraction enabling to transcend the level of sensible experience and reach the world of Ideas. One is immediately reminded of the experiment described in the *Meno*, where the slave boy discovers the existence of truths via geometrical thinking, as in the case of doubling the surface of a square.[2] Geometry is therefore intrinsically linked to the visual way in which concepts are represented and, eventually, drawn, as it is largely shown by Euclid in his *Elements*.[3] The influence of this geometrical thinking goes until the twentieth century, as one may observe by considering, for example, David Hilbert's debt to Euclid as documented in his *Grundlagen der Geometrie* (1899).

[1] This and other very interesting aspects of the role of geometry in Plato's dialogues are developed by Athanase Papadopoulos in his forthcoming essay devoted to the relation between music, mathematics and astronomy starting from Plato's *Timaeus* (Papadopoulos 2020).

[2] There are several editions available of Plato's dialogues, including the two mentioned ones. See, for example, Plato's collected work edited by John M. Cooper and Douglas S. Hutchinson (Plato 1997).

[3] The thirteen books of Euclid's *Elements* are available online at Clay Mathematics Institute Historical Archive (https://www.claymath.org/library/historical/euclid/)

It can be easily shown that this emphasis on the axiomatization of geometry, which is prevalent in Hilbert's work, has a direct influence in the way in which twentieth century composers, starting from Ernst Krenek (1937), approached the problem of the theoretical foundations in music via formalization.[4] A turning point towards a full axiomatic thinking in geometry, with extremely significant musical consequences, is symbolized by the way in which geometry is progressively approached via an algebraic perspective. Starting from Felix Klein's *Erlangen Program* (Klein 1893), the way of conceiving the geometry of a space is in fact intrinsically linked to the way in which transformation groups act on such a space. With the entry of abstract group theory into play we assist to a progressive shift from the purely representational nature of geometrical entities to their truly operational power which has a big influence on the way in which music theorists conceptualize the musical phenomena.[5] This little historical digression may offer some elements to understand the complex link one can establish between the geometrical properties of musical structures and their underlying formalization which accompanied the evolution of mathematical music theory from a simply representational to an operational discipline capable of approaching in a powerful way the dynamic character of musical processes. *Topology* – let us now focus on this term – precisely enters the picture as a subsidiary discipline when the focus on *musical structures* is substituted or at least accompanied by a new attention of music theorists and analysts towards *musical processes*, i.e. theoretical constructions in which time plays a fundamental role. The extensive use of topological techniques in the mathematical modeling of musical structures and processes is a relatively new phenomenon, although one may already find many examples of genuine topological constructions in Guerino Mazzola's *Geometrie der Töne* (1990). The foundational basis of the concept of "global compositions," as opposed to the "local" ones (originally considered as finite non-empty subsets of a module over a ring), is the mathematical structure of Riemannian manifolds allowing the music theorist to make use of several topological constructions. The reader familiar with this approach in mathematical music theory will easily recognize some traditional visualizations of musical structures, such as the Möbius strip[6] and, more generally, the nerve of a *n*-dimensional simplicial complex as representing the complexity of the specific covering that the analyst will associate to a given global composition. Interestingly, there is already in this approach an interplay between geometry and combinatorial topology which shows that there is a shift from the simple characterization of (static) musical structures to (dynamic) musical processes, such as modulation, whose properties are now described in a topological way.[7] More recently, it became clear to music theorists that some of most popular constructions in neo-Riemannian and, more generally, transformational music analysis do intrinsically possess a topological structure. The most celebrated example is probably the *Tonnetz*, originally mostly studied from a graph-theoretical and algebraic perspective, and successively apprehended from a more topological way, enabling the complete characterization of their different instances (or *Tonnetze*) via the study of the associated algebraic invariants, such as Betti numbers (Bigo, Giavitto, and Spicher 2011; Catanzaro 2011; Bigo 2013). This interplay between geometry, algebra and topology provides nowadays a better understanding of the way in which generalized musical spaces can be represented and formalized from a very elegant and

---

[4] See Andreatta (2003) for a first attempt at reconstructing this legacy from both an European and American perspective.

[5] This dialectical principle between the *objectal* and *operational* nature of the reality has been deeply studied by the French philosopher and epistemologist Gilles-Gaston Granger who took this duality as the foundational basis for the very notion of "concept" in philosophy (Granger 1947/1994).

[6] Which the author also calls "Harmonisches Band" to emphasize its truly music-theoretical origin, that he bases on Arnold Schoenberg's writing on the theory of harmony (Schoenberg 1911).

[7] There are other examples that could be considered as genuinely "topological" within the same approach in mathematical music theory. We only mention two of them, both dealing with the formalization of musical motifs as combinatorial and topological structures: Netske (2004) and Buteau and Mazzola (2008).

computationally powerful perspective.[8] Each of the three contributions of the present issue can be considered as offering to the reader a specific perspective on this complex interplay starting from different but intrinsically linked music-theoretical problems.

"Why Topology?" by Dmitri Tymoczko makes use of topological techniques to formalize voice leadings as paths in n-dimensional geometric spaces. The approach is an extension of the conceptual constructions that constitute the foundation of Tymoczko's geometrical theory of music as presented and discussed in details in his book *A Geometry of Music* (Tymoczko 2011). Previous work by the author also uses topological techniques but without the detailed argumentation for the value of topology in contemporary mathematics and music research.[9] Higher dimensional geometric spaces whose points are generic n-notes chords are *orbifolds* in a technical mathematical sense, i.e. quotient spaces obtained by the action of the permutation (or symmetric) group of $n!$ elements on the n-dimensional torus and possess therefore non-trivial topological properties. The present paper focuses on the interpretation of voice-leading properties via the fundamental group and can therefore offer new insights to the analyst to fully understand the author's previous music-theoretical constructions. Moreover, it shows that not only voice leading but, more generally, transformational constructions in music analysis are linked to the fundamental concept of homotopy equivalence, which again puts algebra – and group theory in particular – into play. Although homotopy theory has become very popular in recent years thanks to Guerino Mazzola's gesture theory, the approach described in Tymoczko's paper is fundamentally different since it provides the theoretical background to fully understand the topological properties of different harmonic spaces.

While Tymoczko's voice-leading geometries have led to many significant analytical and music-theoretic applications, as is in evidence of *Geometry of Music*, to some extent there has always been a disconnect between the higher dimensional orbifolds defined in Callender, Quinn, and Tymoczko 2008 and music analysis and composition, because as the spaces get more complex we lose the benefit of direct spatial intuition. In previous work (such as Tymoczko 2011) spatial intuition has often been recovered by simplifying chord spaces to lattices on a limited number of chord-types, which renders their more interesting topological features irrelevant. In the present work, Tymoczko instead uses homotopic equivalence to simplify the spaces. Remarkably, he is able to reduce these topological features to simple two-dimensional representations, making them easily surveyable and available to analytical intuition, regardless of the chord cardinality and resulting complexity of the original space. He introduces two kinds of simplifications: one is the circle of pitch-class sums, which has the topology of an annulus or circle, regardless of cardinality. The homotopy of this space corresponds to total voice-leading ascent and descent. The other simplification is a homotopic classification of set-class spaces (which factor out transposition or transposition or inversion). The analytical applications in Tymoczko's article begin to reveal the wealth of musical interest in the interactions between features of a voice leading that we usually are tempted to "reduce out," such as voice crossings and transpositions along the chord. The article also introduces a principled generalization of neo-Riemannian transformations to any chord type, as voice leadings to its inversions.

[8] We could also add to the picture the categorical perspective which is intrinsically linked to the interplay between geometry and algebra and which can also shed some new light on existing topological approaches not only in music theory and analysis but also in performance studied like in the case of a category and topos-based theory of musical gestures. See in particular the third volume of the revised version of *Topos of Music* (Mazzola 2017). The reader who might be interested in going more deeply into some philosophical and epistemological aspects of the categorical perspectives on the interplay between geometry, algebra, and topology, will find a good starting point in Jean-Pierre Marquis' monograph *From a Geometrical Point of View* (Marquis 2009).

[9] See, in particular, Callender, Quinn, and Tymoczko (2008) for a more detailed presentation of the technical aspects of a geometrical approach in music theory, also including the introduction of orbifold structures into music formalization.

The following paper by Jason Yust, entitled "Generalized *Tonnetze* and *Zeitnetze*, and the Topology of Music Concepts," also focuses on the interplay between the topological structure of the generalized *Tonnetz* seen as a n-dimensional simplicial complex and its possible graph-theoretical and geometric realizations. The notion of a generalized n-chord *Tonnetz* can be articulated in fact at different levels of abstraction, ranging from the traditional presentation as a lattice of chords relating by maximal intersection to more sophisticated representations, embedding this structure on the powerful theory of Fourier phase spaces. This further interpretation adds a new subfield of mathematics to the overall picture, also showing the natural bridges between geometry, algebra, and (mathematical) functional analysis. Many music-theoretical problems, from the construction of rhythmic tiling canons to the classification of homometric musical structures,[10] have been shown to be deeply rooted into Fourier theory, but adequate understanding of the topological implications of such a formalism still constitutes an open research area in mathematics and music. A first crucial achievement has been the characterization of the geometrical structure of generalized *Tonnetze* as toroidal representations in the spaces of Fourier phases, as proposed by Emmanuel Amiot (2013) and further developed by Jason Yust in his critical discussion on the relations between the geometric extensions of Generalized *Tonnetze* and the Fourier phase Spaces (Yust 2018). The present article pushes the previous conclusions further by focusing on the implications of geometrical *Tonnetze* interpreted as embedding a simplicial *Tonnetze* in some geometric space. Yust's article complements Tymoczko's in showing how a different theoretical emphasis, on shared pitch-class content between chords rather than the distances moved by voices, leads in the direction of different kinds of topologies. Despite the different mathematic foundations of their spaces (Fourier phase vs. voice-leading orbifold), the two theories arrive at surprisingly similar constructions in places, with some of Yust's foldings producing spaces topologically equivalent to Tymoczko's pitch-class sum annuli. Despite appearances however, a closer examination reveals that a path or homotopy-class of paths in the two kinds of space actually have distinct musical meanings. The two approaches also lead to interestingly different generalizations of neo-Riemannian transformations, with Tymoczko focusing on the idea of inversional voice leadings while Yust emphasizes maximal intersection.

There is much music-theoretical and mathematical potential in using this new geometric approach, as the reader will see by following Yust's systematic presentation of a panoply of generated collections. These include dyadic *Tonnetze* (that reduce to the traditional chromatic and fifths-generated circular representations of the equal-tempered system), trichordal *Tonnetze* and higher-dimensional *Tonnetze*. All these geometric structures are first defined in toroidal spaces before applying higher-dimensional foldings which shed a new light on different n-chord generated spaces. To this family belong, for example, the *Tonnetz* of chromatic tetrachords and of different 4-note chords and also *Tonnetze* of ninth chords, which are rare in the literature. Special attention is given to *Tonnetze* of diminished seventh chords, which enables the author to stress the fact that an object that is trivial from a graph-theoretical and simplicial perspective can still be interesting from a geometrical and music-theoretical perspective. In addition to its elegant and powerful character, taking phase spaces as a geometrical realization of *Tonnetze* has dramatic consequences in the way in which traditional neo-Riemannian harmonic constructions can be generalized and applied to other musical parameters, and in particular to rhythm. Building bridges between harmonic and rhythmic spaces has a long tradition in music theory and composition, as the history of serialism clearly shows. Many other attempts at providing a general theory on the structural relations between harmonic and rhythmic domains have marked the development of mathematical music theory, from Mazzola's formalization of the two domains as local/global compositions based on the same underlying module structure to more perceptual-based approaches, such as Jeff Pressing's foundational studies on the cognitive isomorphism

[10] Summarized in Amiot (2016).

between pitch and rhythm in ethnomusicology (Pressing 1983). In defining a "*Zeitnetz*" in analogy with the *Tonnetz*, the paper also suggests new very interesting connections between research in mathematical music theory and problems raised by scholars working on different areas, in particular cognitive and empirical musicology.

The final paper by Mattia G. Bergomi and Adriano Baratè, entitled "Homological persistence in time series: an application to music classification", provides a third perspective on the interplay between geometry, topology and algebra focusing on the problem of automatic stylistic music analysis. Based on persistence homology as a specific domain in the field of topological data analysis, the paper shows not only the possibility but, for some extend, the necessity of establishing a fruitful dialogue between two apparently orthogonal research tradition in the field of application of mathematical models to music analysis: mathematical music theory and music information research (MIR). Despite efforts to bring together researchers belonging to the *Society for Mathematics and Computation in Music* with those working in the area of area of the *Society for Music Information Retrieval*, it is not difficult to recognize that there are deep methodological and theoretical differences between these two major orientations in the field of computational musicology.[11] The paper by Bergomi and Baratè begins to fill the gap by applying a panoply of topological, geometrical, and algebraic techniques to the domain of automatic stylistic music classification. Providing models and tools for the automatic stylistic classification is probably one of the most active areas in the field of music information research. Moreover, the authors' focus on time series suggests that topology not only applies to the formalization of musical structures but clearly deals with the modeling of musical processes (or, what they call "time-varying systems"). Using persistence homology as the main conceptual and theoretical framework for approaching style analysis is a relatively recent and very promising approach. The approach originates in one of the authors' dissertation (Bergomi 2015) which gave rise to different research directions exploring the interplay between geometry, topology, and, more recently, category theory.[12] The present paper clearly shows the interplay between static topological representations of musical structures and dynamic geometric formalizations of these topological features. In the former case, the authors use persistent homology as the main topological framework to represent the inner structure of musical object via a collection of persistent diagrams. These static representations are then studied dynamically by formalizing their temporal evolutions as time series that can be compared through dynamic time warping. This very rich mixture of topological and geometrical techniques in the representation and formalization of musical structures and processes, together with their implementation in computer-aided analytical systems, also enable the authors to offer a new and complementary computational perspective on the generalized *Tonnetz* with respect to the one presented in the first two articles. It also starts with the same simplicial complex interpretation of the *Tonnetz* as described in Yust's article, but instead of focusing on their subcomplexes generated by a sequence of pitch classes (as extensively studied in Bigo 2013 and Bigo et al. 2013) it proposes to take into account additional music information in order to

[11] As pointed out by Anja Volk and Aline Honingh in their introduction of the special issue of the Journal of Mathematics and Music devoted to "Mathematical and computational approaches to music" (Volk and Honingh 2012), this gap was already acknowledged by Thomas Noll and Robert Peck in the first issue of the Journal of Mathematics and Music in 2007 (Noll and Peck 2007).

[12] See, in particular, Bergomi, Baratè, and Di Fabio (2016) for a topological perspective and Bergomi et al. (2019) for the first attempts at going beyond a topological approach in the study of a category theory-based persistent homology. Some extensions of the original framework have also been intensively studied within the SMIR (Structural Music Information Research) project. This research project is a permanent interdisciplinary research axis supported by CNRS and led by Moreno Andreatta at the University of Strasbourg / IRMA (Institut de Recherche Mathématique Avancée), in a formal collaboration with IRCAM. Research topics include Mathematical Morphology, Formal Concept Analysis, Persistent Homology and automatic classification of musical styles, Category theory and transformational (computer-aided) music analysis as well as the study of musically-driven open conjectures in mathematics and the epistemological and cognitive implications of contemporary mathemusical research. See Andreatta (2018) for a short overview of the different research topics included in the SMIR project.

break the isotropic character of the simplicial complex. The new geometric space in which musical structures and processes are represented is therefore a deformed *Tonnetz*, i.e. a non-isotropic *Tonnetz* exhibiting variable geometric features which can be retrieved via ad-hoc filtration functions. This enables the authors to approach automatic stylistic music classification in a geometric structural way, which is very original in the field of music information research, where statistics and machine learning constitute the main theoretical framework for retrieval purposes. The tests that have been performed by the author on different data sets, and which are discussed in the final part of the article, show very promising results that shed new lights on the nature of musical style.

We hope that this short introduction provides the reader with enough elements to appreciate not only the singular character of each of the three papers published in this special issue but their deep and fruitful intersections. We insisted several times on the necessity of building bridges between different music-theoretical and computational approaches; but before establishing connections between research groups having different aims and methodological criteria, it is important to reinforce the synergies within the members of the same community. Research carried out by members of the Society for Mathematics and Computation in Music has spread in a multitude of different directions, as the readers of our Journal can easily see. Whereas each of the three contributions offers distinct perspectives on the same music-theoretical constructions, all of them together provide a comprehensive picture of the interplay between geometry and topology which is surely more than the sum of the three local perspectives. Probably the time for a synthetic view on all these different approaches has not yet come, but the reader has the opportunity to build his or her own conceptual bridges between geometry and topology and their mutual contribution to music representation and formalization.

Disclosure statement

No potential conflict of interest was reported by the author(s).

References

Amiot, Emmanuel. 2013. "The Torii of Phases." In *Mathematics and Computation in Music. MCM 2013*, edited by Jason Yust, Jonathan Wild, and John Ashley Burgoyne, 1–18. Berlin, Heidelberg: Springer.

Amiot, Emmanuel. 2016. *Music through Fourier Space: Discrete Fourier Transform in Music Theory.* Heidelberg: Springer.

Andreatta, Moreno. 2003. "Méthodes algébriques dans la musique et musicologie du XXe siècle: aspects théoriques, analytiques et compositionnels." PhD thesis, Ecole des hautes études en sciences sociales, IRCAM, Paris.

Andreatta, Moreno. 2018. "From Music to Mathematics and Backwards: Introducing Algebra, Topology and Category Theory into Computational Musicology." In *Imagine Math 6 - Mathematics and Culture*, edited by M. Emmer, and M. Abate, 77–88. Berlin: Springer.

Bergomi, Mattia G. 2015. "Dynamical and Topological Tools for (Modern) Music Analysis." Ph.D. thesis, Université Pierre et Marie Curie, Paris.

Bergomi, Mattia G., Adriano Baratè, and Barbara Di Fabio. 2016. "Towards a Topological Fingerprint of Music." In *Topology in Image Context*, edited by A. Bac and J.-L. Mari, 88–100. Switzerland: Springer international Publishing.

Bergomi, Mattia G., Massimo Ferri, Pietro Vertechi, and Lorenzo Zuffi. 2019. "Beyond Topological Persistence: Starting from Networks." *arXiv* preprint https://arxiv.org/abs/1901.08051.

Bigo, Louis. 2013. "Représentation symboliques musicales et calcul spatial." PhD Thesis, université Paris Est Créteil / IRCAM.

Bigo, Louis, Moreno Andreatta, Jean-Louis Giavitto, Olivier Michel, and Antoine Spicher. 2013. "Computation and Visualization of Musical Structures in Chord-Based Simplicial Complexes." In *Mathematics and Computation in Music, MCM 2013*, edited by Jason Yust, Jonathan Wild, and John Ashley Burgoyne, 38–51. Heidelberg: Springer.

Bigo, Louis, Jean-Louis Giavitto, and Antoine Spicher. 2011. "Building Topological Spaces for Musical Objects." In *Mathematics and Computation in Music. MCM 2011*, edited by Carlos Agon, Moreno Andreatta, Gérard Assayag, Emmanuel Amiot, Jean Bresson, and John Mandereau, 13–28. Berlin, Heidelberg: Springer.

Buteau, C., and G. Mazzola. 2008. "Motivic Analysis According to Rudolph Réti: Formalization by a Topological Model." *Journal of Mathematics and Music* 2 (3): 117–134.

Callender, Clifton, Ian Quinn, and Dmitri Tymoczko. 2008. "Generalized Voice-Leading Spaces." *Science* 320 (5874): 346–348.

Catanzaro, Michael. 2011. "Generalized *Tonnetze*." *Journal of Mathematics and Music* 5 (2): 117–139.

Granger, Gilles-Gaston. 1947. "Pygmalion. Réflexions sur la pensée formelle." *Revue Philosophique* (7–9): 282–300. Reprinted in *Formes, opérations, objets*. Paris: Librairie Philosophique J. Vrin, 1994.

Hilbert, David. 1899. *Grundlagen der Geometrie*. Stuttgart: Teubner.

Klein, Felix. 1893. "A Comparative Review of Recent Researches in Geometry." *Bulletin of the American Mathematical Society* 2 (10): 215–250.

Krenek, Ernst. 1937. *Über Neue Musik. Sechs Vorlesungen zur Einführung in die theoretischen Grundlagen.* Vienna: Universal Edition. Wien (English translation as *Music here and now*, New York. Norton, 1939).

Marquis, Jean-Pierre. 2009. *From a Geometrical Point of View. A Study of the History and Philosophy of Category Theory*. Berlin, Heidelberg: Springer.

Mazzola, Guerino. 1990. *Geometrie der Toene*. Zürich: Birkhäuser.

Mazzola, Guerino, R. Guitart, J. Ho, A. Lubet, M. Mannone, M. Rahaim, and F. Thalmann 2017. *The Topos of Music III: Gestures. Musical Multiverse Ontologies*. Berlin, Heidelberg: Springer.

Netske, Andreas. 2004. "Paradigmatic Motivic Analysis." In *Perspectives in Mathematical and Computational Music Theory*, edited by Guerino Mazzola, Thomas Noll, and Emilio Lluis-Puebla, 343–365. Osnabrück: Osnabrück Series on Music and Computation.

Noll, Thomas, and Robert Peck. 2007. "Welcome." *Journal of Mathematics and Music* 1 (1): 1–6.

Papadopoulos, Athanase. 2020. "Mathématiques, musique et cosmologie: à partir du *Timée* de Platon." forthcoming in *Les jeux subtils de la poétique, des nombres et de la philosophie : Autour de la musique de Walter Zimmermann*, edited by Pierre Michel, Moreno Andreatta, and José Luis Besada. Paris: Hermann.

Plato. 1997. *Complete Works*. Edited by John M. Cooper and Douglas S. Hutchinson. Indianapolis/Cambridge: Hackett Publishing Company.

Pressing, Jeff. 1983. "Cognitive Isomorphisms Between Pitch and Rhythm in World Musics: West Africa, the Balkans, and Western Tonality." *Studies in Music*, 17: 38–61.

Schoenberg, Arnold. 1911. *Harmonielehre*. Vienna: Universal Edition.

Tymoczko, Dmitri. 2011. *Geometry of Music: Harmony and Counterpoint in the Extended Common Practice*. New York: Oxford University Press.

Volk, Anja, and Aline Honingh. 2012. "Mathematical and Computational Approaches to Music: Challenges in an Interdisciplinary Enterprise." *Journal of Mathematics and Music* 6 (2): 73–81.

Yust, Jason. 2018. "Geometric Generalizations of the Tonnetz and their Relation to Fourier Phase Spaces." In *Mathematical Music Theory: Algebraic, Combinatorial, Topological and Applied Approaches to Understanding Musical Phenomena*, edited by Mariana Montiel, and Robert Peck, 253–278. Singapore: World Scientific.

Why Topology?

Dmitri Tymoczko

Music theorists have modeled voice leadings as paths through higher-dimensional configuration spaces. This paper uses topological techniques to construct two-dimensional diagrams capturing these spaces' most important features. The goal is to enrich set theory's contrapuntal power by simplifying the description of its geometry. Along the way, I connect homotopy theory to "transformational theory," show how set-class space generalizes the neo-Riemannian transformations, extend the *Tonnetz* to arbitrary chords, and develop a simple contrapuntal "alphabet" for describing voice leadings. I mention several compositional applications and analyze short excerpts from Gesualdo, Mozart, Wagner, Stravinsky, Schoenberg, Schnittke, and Mahanthappa.

Geometrical music theory models n-voice chords as points in a generalized donut, a twisted, mirrored, higher-dimensional orbifold that is difficult to visualize and understand. Some theorists have simply recorded how music moves through these spaces, relying on the reader's visual intuition to convert geometrical paths into analytically useful observations. Others have described parts of the geometry, typically graphing single-step voice leading among the nearly even chords in some particular scale (Douthett and Steinbach 1998; Tymoczko 2011, §3.11). This restriction has been justified on the grounds that maximally even sonorities are optimal from various points of view (e.g. combining harmonic consistency with stepwise voice leading). But we do not have to look hard to find successful music departing from optimality: more than fifty years ago Herbert Simon criticized the tendency to overemphasize the superlative, coining the term "satisficing" to refer to the pursuit of the merely good-enough (Simon 1956). Hence the need for voice-leading models applying to *arbitrary* chords and scales – including those that might not be theoretically ideal.

This paper will use topology for this purpose, proposing simple models of the voice-leading relations among arbitrary chords and set classes. The paper's main innovation is to associate structural features of existing geometrical spaces (such as "loops" and "boundaries") with familiar voice-leading transformations (such as "transpositions along the chord," "voice exchanges," and "generalized neo-Riemannian transformations"). This leads to two-dimensional representations of voice leading that are structure-neutral, applying to any chord in any scale and hence freeing us from the constraints of near-evenness and single-step voice leading. These representations simultaneously describe the topology of the higher-dimensional voice-leading spaces and the algebraic structure of the voice leadings connecting chords, with transformational theory providing a tool for exploring the geometry of contrapuntal possibility. This in turn will lead me to distinguish two different sorts of musical transformation, global symmetries that show us how to fold up musical space and the contrapuntal possibilities that describe how we can move around the resulting geometry.

My argument is long and detailed, so it may help to keep the following picture in mind. Musical symmetries create configuration spaces whose features correspond to different kinds of voice leading; attending to some subset of voice-leading possibilities leads to simplified models containing only the relevant geometrical features. To understand the geometry it is therefore useful to understand voice-leading transformations and vice versa. The (bijective) voice leadings from a chord to any of its transpositions can be decomposed into a sequence of simpler voice leadings VTR, where V is a (possibly complicated) voice exchange, T is a transposition along the scale moving every note in the same direction by the same number of steps, and R is a "residual" to be described momentarily.[1] These voice leadings can be associated with elements in what is called the "fundamental group" of voice-leading space, recording the various kinds of loops to be found in the n-dimensional mirrored donut modeling n-note chords; intuitively the voice crossings V correspond to the mirrored boundaries of chord space while the transpositions T are associated with its circular dimension. The voice exchanges and transpositional voice leadings generate normal subgroups which can be safely ignored (abstracted-away-from or "quotiented out"). Deemphasizing voice exchanges is broadly familiar from Schenkerian analysis and characteristic of the two-dimensional annular spaces in §2 below; it corresponds to disregarding the singular, mirrored boundaries of chord space.[2] Deemphasizing transposition is familiar from set theory and embodied in the two-dimensional polygonal diagrams of §3–4; it corresponds to disregarding the circular dimension of chord space. In transpositional set-class space, the "residual" voice leadings are "transpositions along the chord" to be discussed below. In inversional set-class space, the residual voice leadings generalize the familiar transformations of neo-Riemannian theory. Topology thus connects diverse music-theoretical fields from Schenkerian analysis to set theory to neo-Riemannian theory.

One practical benefit of these ideas is to clarify the music-theoretical notion of *inversion*. As we will see in §§2–3, the residual transpositions along the chord are forms of *registral inversion,* described by terms like "root position," "first inversion," and "second inversion." But where traditional theory is concerned only with a chord's lowest note, constructing inversions by successively moving bass to soprano, the geometrical approach preserves chordal spacing as measured in chordal steps; this is useful in contexts where spacing is musically significant. The residual transformations of inversional set-class space (§4) extend this idea to what set theorists call *pitch-* or *pitch-class* inversion, an operation that turns musical space upside-down; these generalizations of the familiar neo-Riemannian transformations again preserve spacing as measured in chordal steps. As we will see, the residual transformations operate similarly on both *notes* and *voices*: a "transposition along the chord" sends note x in voice y to note $x + c$ in voice $y + d$, while a generalized neo-Riemannian inversion sends note x in voice y to note $c–x$ in voice $d–y$ (assuming voices labeled in pitch-class order). It is this double action that allows for efficient voice leading, with the operation on voices negating or nearly negating the operation on pitch classes.

Since my goal is to introduce a general framework to a broad musical audience, I try to write in a way that is comprehensible to nonmathematicians, presuming only a basic understanding of

[1] "Bijective" voice leadings between non-transpositionally related chords can involve doublings but those must be fixed in advance. Nonbijective voice leadings, in which doubling is more flexible, are more complex but can be treated using the tools introduced here; exploring this is a topic for future work. The term "scale" includes the limiting case of continuous and unquantized pitch space. See the appendix for more.

[2] Here I am referring to the general Schenkerian idea that voice exchanges decorate a simpler uncrossed background (e.g. Brown 2005, 78).

voice-leading geometry. For the same reason, I do not offer lemmas, theorems, and proofs, trusting that mathematically inclined readers can either find these in the existing literature or supply them themselves. (I do, however, offer mathematical details in the appendix.) Each section summarizes the central concepts before diving into the technical weeds. I have also created five websites allowing users to explore these ideas in a hands-on way; I strongly encourage readers to use these interactive resources as they work through the paper, for beyond understanding general principles, music theorists need a good intuitive sense for how to make their way around the spaces they study.[3]

1. Mysteries of the *Tonnetz*

For centuries people have been drawing pictures of musical possibilities. One of the oldest, the *Tonnetz*, originates with the eighteenth-century mathematician Leonhard Euler, who used the figure to show how major and minor chords were composed of consonant intervals. With perfect fifths on the horizontal axis and major thirds on the northwest diagonal, major and minor triads can be represented by equilateral triangles (Figure 1).[4] Two chords share a vertex when they have one common tone and a side when they have two. Mathematically, the figure is a *torus* because its left edge is identified with its right and its bottom edge is identified with its top: on Figure 1, the left side D♭–F–A is duplicated as F–A–C♯ on the right, and the bottom A–E–B–F♯–C♯ is respelled on the top as B♭♭–F♭–C♭–G♭–D♭. Thus as we move off one edge we reappear on the opposite side, much like an old-fashioned video game.

More than two centuries later, Richard Cohn (1996) noticed that the *Tonnetz* could be repurposed to represent *voice leading* among major and minor triads.[5] He used "triangle flips" to model progressions with two common tones and one stepwise melodic voice, with the triangle's three vertices representing three musical voices. These flips, illustrated in Figure 2, evoke relations central to Hugo Riemann's music theory: the *parallel* relationship between triads sharing a perfect fifth; the *relative* relationship between triads sharing a major third; and the *leading-tone-exchange* between triads sharing a minor third. Reinterpreted as voice-leading transformations, or things you can do to a chord, these triangle-flips combine to generate a wide range of contrapuntal possibilities, providing a useful tool for modeling complex chromatic passages. The result is an intriguing connection between eighteenth-century mathematics, nineteenth-century theory, and twentieth-century analysis.

The reinterpreted *Tonnetz,* however, is more subtle than it seems. The two paths in Figure 3 look similar, both moving along one of the figure's "alleys" to return to their starting point by a series of triangle flips. Yet when we compare the resulting music we find that the southeast path returns every voice to its initial position while the northeast path moves each voice down to the next chord tone: G down to E, E down to C, and C down to G. The cumulative effect of the southeast path is *trivial*, equivalent to doing nothing at all, whereas the cumulative effect of the northeast path is a nontrivial "transposition along the chord." This difference is suggestive in light of the mathematical discipline known as homotopy theory, where trivial loops are precisely those that can be smoothly contracted to a point. Perhaps only one of the *Tonnetz*'s axes surrounds a genuine hole in the space of contrapuntal possibilities.

---

[3] See http://dmitri.mycpanel.princeton.edu/cs.html and /multichord.html for the annular spaces of §2, /sc.html for the set-class polygons of §§3–4, /sc.html and /nr.html for the voice-leading alphabet in §5, and /nr.html and /tonnetz.html for generalized neo-Riemannian theory (§4). These programs, along with explanatory videos, can also be found at madmusicalscience.com.

[4] For histories of the *Tonnetz* see Cohn (1997, 2011b). The specific version I have used, with tilted rather than rectangular axes, originates with Otakar Hostinský.

[5] Here Cohn drew on the work of David Lewin (1987, 175ff) and Brian Hyer (1989), both of whom understood neo-Riemannian transformations as contextual inversions applying to entire chords. See §4 for more.

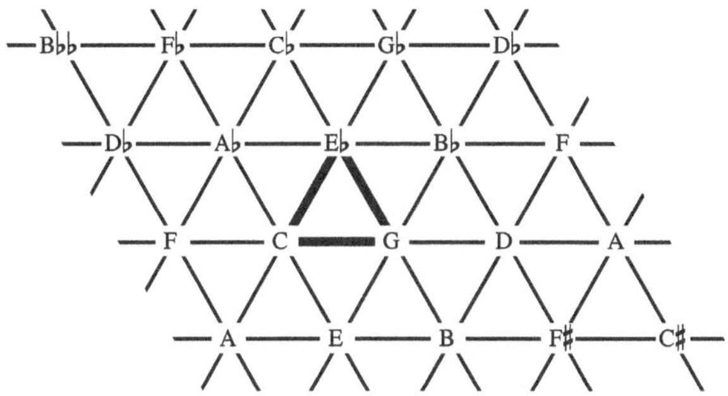

Figure 1. The *Tonnetz*.

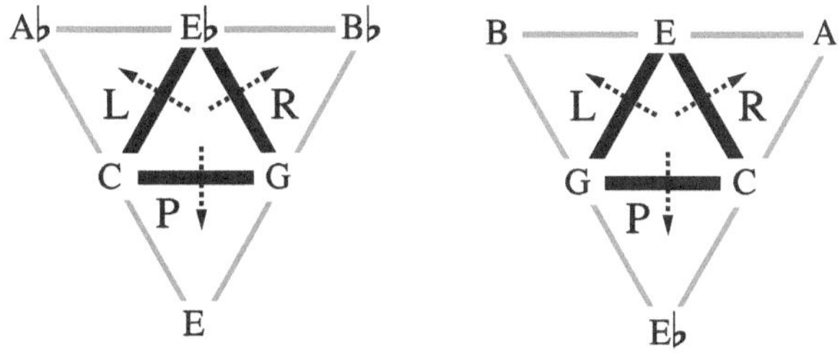

Figure 2. L, P, and R triangle flips for minor triads (left) and major triads (right). Each flip fixes an edge whose vertices represent common tones; the third voice moves by one or two semitones.

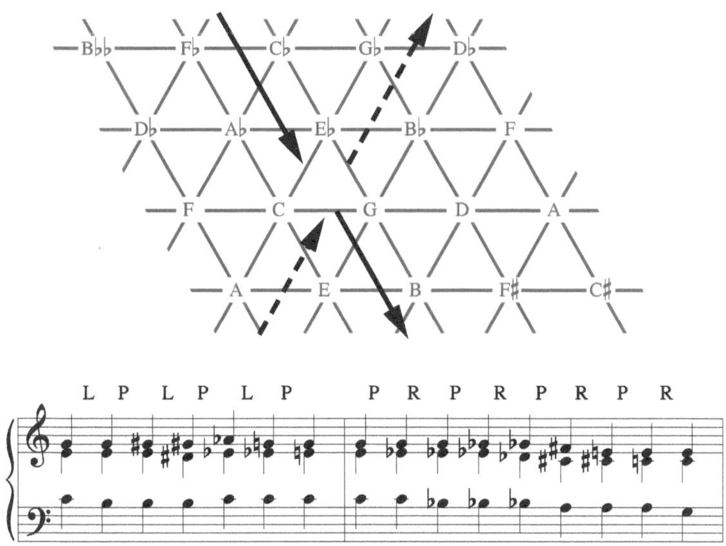

Figure 3. The solid path represents the LP cycle (LPLPLP) and returns each note to its starting point; the dashed path represents the PR cycle (PRPRPRPR) and moves each voice downward along the C major chord.

This phenomenon is surprisingly general. Figure 4 shows F. G. Vial's eighteenth-century model of key relations, made popular by Gottfried Weber. Here points represent keys with modulations along the SW/NE diagonal moving by fifth and modulations along the SE/NW diagonal

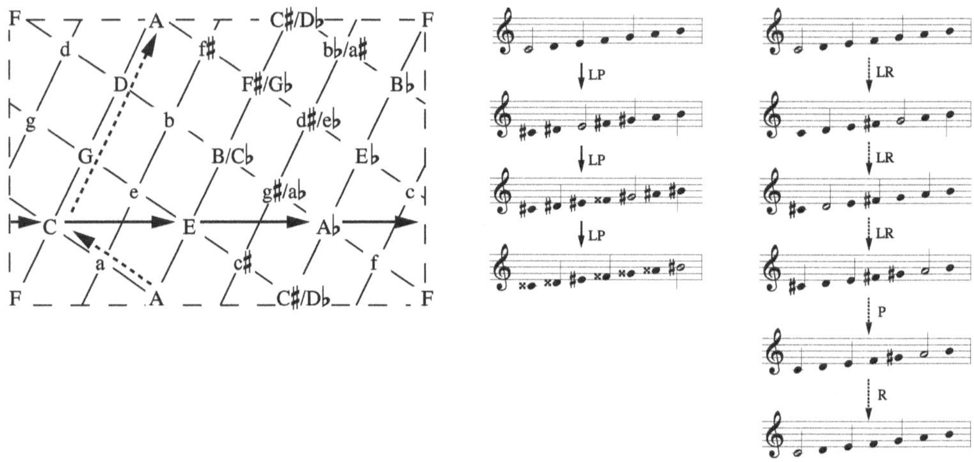

Figure 4. Two paths on the chart of the regions, again with different effects.

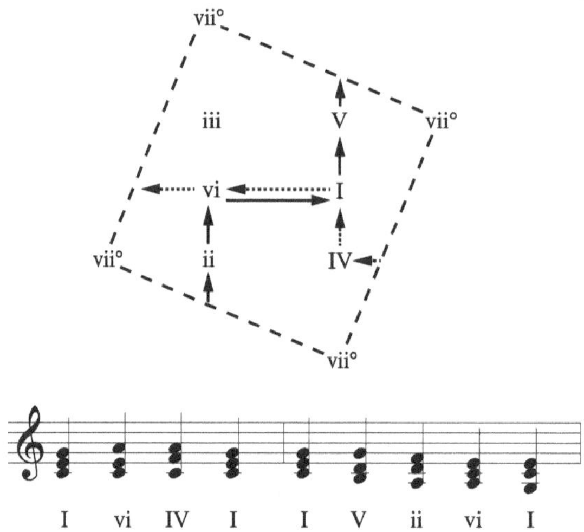

Figure 5. Lerdahl's toroidal model of diatonic triads, along with two paths in the space.

alternating between "parallel" and "relative" keys. These can be formalized as applying separate voice leadings to a tonic triad and a superordinate scale as described in chapter 4 of Tymoczko (2011).[6] The figure shows two similar-seeming paths that loop around different axes. As before, one leaves each note of the scale where it began while the other shifts each scale degree up by step. However, the location of the trivial loops has shifted: on the triadic *Tonnetz* a series of LP moves forms a trivial loop (Figure 3), while on Vial's figure it has a nontrivial effect on the scale.[7] Thus even though the figures are visually similar, they represent distinct spaces of contrapuntal

[6] On Vial's map the L transform applies L to the tonic triad while adding or subtracting one of the scale's sharps; P leaves the tonic triad unchanged but adds or subtracts three of the scale's sharps; and R applies R to the tonic triad while leaving the scale unchanged.

[7] See Cohn (2011b) for a perspective that emphasizes the similarity between the traditional *Tonnetz* and Vial's figure.

possibility. Figure 5 identifies a similar issue in Fred Lerdahl's map of the diatonic triads.[8] From a superficial perspective these graphs have the topology of a two-dimensional torus, but when we treat them as voice-leading spaces we find hints of something else.

Cohn's reinterpreted *Tonnetz* was a significant music-theoretical achievement, a quantitative and geometrical model of the voice leadings connecting major and minor triads. But its structure remains poorly understood: while theorists long ago noticed the difference between "trivial" and "nontrivial" paths on the *Tonnetz*, they have struggled to provide a principled explanation for the difference; nor is it obvious how triangle flips relate to the more general phenomenon of efficient voice leading, nor how to extend the *Tonnetz* to larger collections.[9] We will answer these questions over the course of the following pages.

2. Topological models of chord space

How can we understand the circular topology that so often appears in our voice-leading models? This section will answer that question by describing two-dimensional annular spaces that are topological in a double sense: first, in being neutral with respect to chord structure, and second in associating voice leadings with *homotopy classes* of paths.[10] These annular structures are most useful when we ignore voice exchanges to focus on "strongly crossing-free" voice leadings, but we will see that they can be extended beyond this limitation.

I begin by reviewing some basic definitions. A *path in pitch-class space* is an ordered pair (p, r) whose first element is a pitch class and whose second element is a real number representing how the note moves, measured in scale steps, with positive numbers corresponding to ascending musical motion and negative numbers descending. In this formalism pitch classes are subject to octave equivalence, but musical motions are not: I distinguish ascents by x, $x + \boldsymbol{o}$, $x + 2\boldsymbol{o}$, etc., as well as descents by $\boldsymbol{o}–x$, $\boldsymbol{o}–x–\boldsymbol{o}$, $\boldsymbol{o}–x–2\boldsymbol{o}$, etc. (with $\boldsymbol{o}$ the size of the octave and $0 \leq x < \boldsymbol{o}$). These represent musical actions like "start at any C and move up by 4 semitones," "start at any C and move up by 16 semitones," and "start at any C and move down by 8 semitones," all of which are aurally distinct, even though they all move C to E.

A *voice leading* is a multiset of paths in pitch-class space, representing a specific way of moving the notes of one chord to those of another: "C stays fixed, E moves up by one semitone, G moves up by two semitones." (*A transpositional voice-leading schema* is an equivalence class of voice leadings related by transposition: "the root of the major triad stays fixed, the third moves up by semitone, the fifth moves up by two semitones."[11]) A voice leading is *strongly crossing free* if its voices never cross no matter how they are arranged in register. The space of n-note chords is a singular quotient space, the orbifold $\mathbb{T}^n/\mathbf{S}_n$. We form it by taking the n-dimensional torus, whose axes are all circular, and gluing together those points whose coordinates are equivalent under reordering. The result is a twisted higher-dimensional mirrored donut: a space with one circular dimension, whose coordinate is the sum of its chords' pitch classes, with the remaining dimensions forming a simplex or higher-dimensional triangle; the space is bounded by mirror singularities representing chords with duplicate pitch classes. Paths in these spaces can be associated with voice leadings. A fundamental idea of voice-leading geometry is, first, to associate

[8] Lerdahl (2001, 2020) does not use this space to model voice leading, but rather an abstract harmonic distance that serves as one input to a complex algorithm whose ultimate output is a "tension value," one of the primary empirical quantities in his theory. My point is that if we *were* to use the "diatonic torus" to model voice leading, we would find the same issue once again.

[9] See Siciliano (2002, 134), for comments on trivial and non-trivial loops.

[10] Two loops are homotopic, or belong to the same homotopy class, if one can be smoothly deformed into the other. In the musical context this means that the voice leading corresponding to a path depends not on its specific geometry but on its general topological features.

[11] See the appendix for a more formal definition.

these paths with the glissandi they naturally represent, and second, to associate these glissandi with the *discrete* voice leadings to which they are formally equivalent. This is not because actual music involves actual glissandi, but rather because the two different kinds of object have the same mathematical structure. (Indeed, this link between the discrete and the continuous is characteristic of homotopy theory, and voice leadings as I define them are isomorphic to homotopy classes of paths in these configuration spaces.) Readers unfamiliar with these ideas can review the appendix or see Tymoczko (2011) for a thorough introduction.

Geometrical music theory describes higher-dimensional spaces containing *all* possible voice leadings among *all* possible n-note chords. The models in this section use two dimensions to depict *some* of the voice leadings among *some* n-note chords. The basic idea is to place a collection of n-note chords, useful for some particular analytical or theoretical purpose, in an *annulus* or punctured disk, with angular position given by the sum of the chord's pitch classes modulo the octave. As in the full geometrical spaces, points represent chords and continuous paths represent voice leadings between their endpoints. The circular dimension of the annulus corresponds to the circular dimension of the higher-dimensional spaces, with the remaining n–1 dimensions compressed into a line segment.[12] The boundaries of the annulus, like those of higher-dimensional chord space, represent voice exchanges, and I will avoid placing chords there. Chords' radial positions are otherwise completely free, and indeed not fundamental to the model – though in my experience it is useful to place distinct chords at different spatial locations; after all, part of the point of geometrical modeling is to provide intuitive pictures of musical phenomena. What is surprising is that relatively little analytical power is lost by these simplifications.

Figure 6 draws the annular spaces for the standard chromatic and diatonic scales, in the latter case treating diatonic steps as having size 1. I arrange pitch-class sums so that clockwise motion is descending; this is because descending musical motion and clockwise geometrical motion are both defaults. When I am modeling a collection of transpositionally related chords, I will sometimes place them along a spiral representing transposition as in Figure 19 below, but this is merely to facilitate intuition: all relevant calculations can be performed without the spiral. These models are not consistent if we try to combine chords of different cardinalities, for reasons discussed in Callender, Quinn, and Tymoczko (2008) and Genuys (2019).[13]

Voice leadings that touch the boundaries (either of annular space or the higher-dimensional spaces they represent) are those with voice exchanges, or *voice crossings in pitch-class space;* in this section, we will mostly ignore them to focus on bijective *strongly crossing-free voice leadings,* or those that preserve a chord's registral ordering no matter how its notes are deployed in pitch space.[14] (Again, this deemphasizing of voice exchanges is reminiscent of Schenkerian thinking.) These are represented by paths, not touching any boundary, whose angular component is determined by the *sum of the real numbers in the voice leading's paths* (Figure 7). Since there is at most one bijective strongly crossing-free voice leading for every real number, every sequence of voice leadings that forms a complete circle, not enclosing the center of the annulus, composes to form a "trivial" voice leading leaving every voice exactly where it began; by contrast, a complete circle enclosing the center represents a *strongly crossing-free voice leading from a chord to itself involving one octave of total motion* (Figure 8). These motions are sometimes known as "transpositions along the chord," moving each voice along the chord as if it was a very

---

[12] A simplex is the higher-dimensional analogue of a triangle or tetrahedron, an n-dimensional figure bounded by $n+1$ points all connected by edges. Topologically we can think of it as a closed ball. See Callender, Quinn, and Tymoczko 2008; Tymoczko 2011.

[13] Intuitively, the issue is that transposition by **o**/n preserves a chord's radial position, preventing us from embodying transpositional symmetry when we try to combine chords of different sizes

[14] See the appendix or Tymoczko (2011) for technical details. While we can think of the two boundaries of the annulus as representing voice exchanges or voice crossings, there is no principled way to associate them with particular voice exchanges.

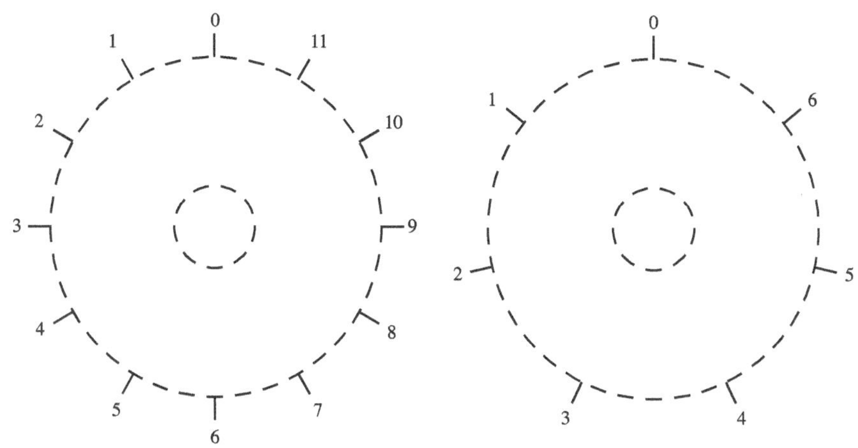

Figure 6. Annular models of chromatic and diatonic space. Angular position corresponds to pitch-class sum.

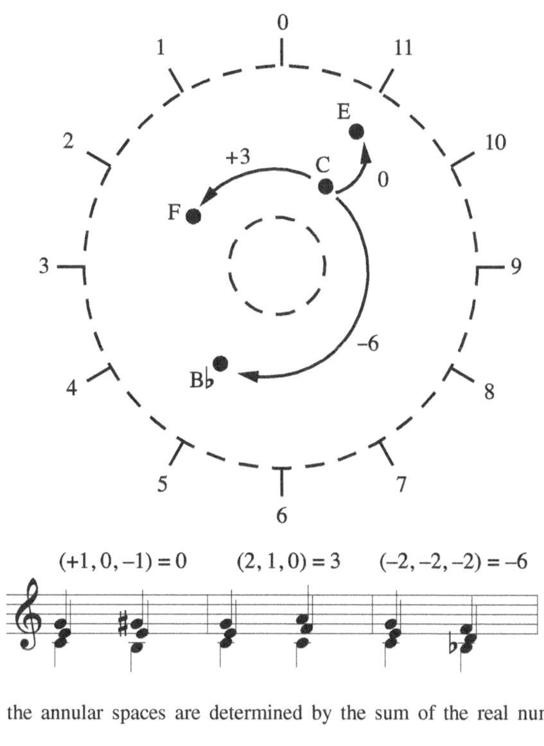

Figure 7. Trajectories in the annular spaces are determined by the sum of the real numbers in a voice leading's pitch-class paths.

small scale; they can be conceived as the higher-dimensional analogues of octave shifts. I will notate them with a lowercase t (see the appendix for a list of notational conventions). The (bijective) strongly crossing-free voice leadings from a chord to itself $X \to X$ are transpositions along the chord t_i; those connecting distinct transpositions combine transpositions along the chord t_i with transpositions along the scale T_j. The voice leadings t_n and T_o are equivalent both musically and topologically, moving every voice by octave and making n loops through chord space.

For chords unrelated by transposition we can use a little algebra. Let $\boldsymbol{o}$ be the size of the octave and let X and Y be a pair of chords whose pitch classes sum to ΣX and ΣY; then for every

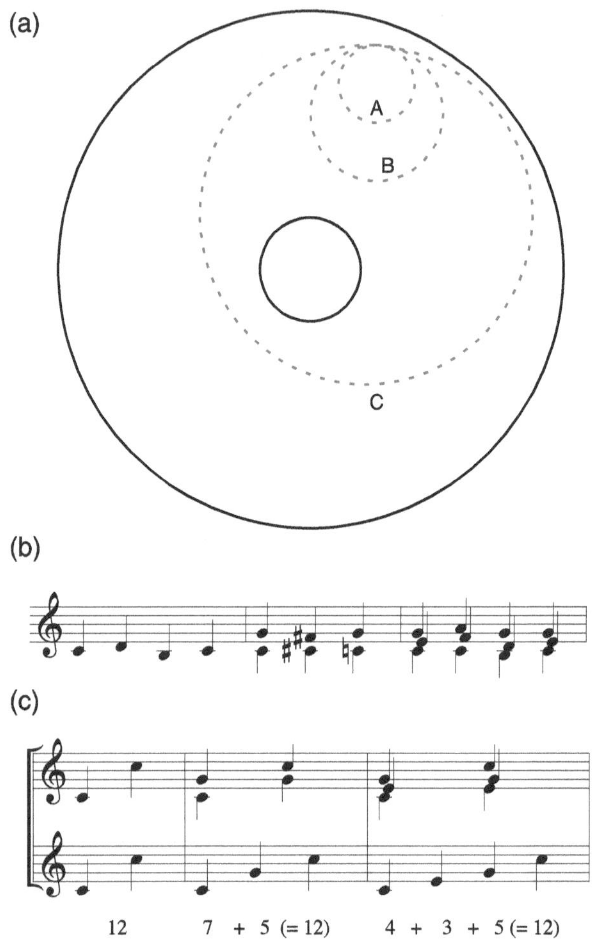

Figure 8. Loops such as A and B, which do not enclose the center of the space, return each voice to its starting point, as in the passages in (b). Loops such as C, which do enclose the center of the space, represent "transpositions along the chord" in which each note moves up or down by one chordal step (c). The top staff of (c) shows the voice leadings while the bottom staff shows that they each involve an octave of total motion in all voices.

number $s = \Sigma Y - \Sigma X + io$, with integer i, there is exactly one strongly crossing-free voice leading $X \rightarrow Y$ whose paths' real numbers sum to s. We can derive these by composing an arbitrary strongly crossing-free voice leading $X \rightarrow Y$ with the set of transpositions-along-the-chord $Y \rightarrow Y$ (Figure 9). (Musically we can identify the resulting voice leadings by placing both X and Y in "close position" so that they span less than an octave; moving the bottom note of one chord to the top, or top to bottom, while continuing to map notes in registral order, changes the total sum of the real numbers in the voice leading's paths by one octave.) Here, the integer i determines how many times the voice leading $Y \rightarrow Y$ circles the center of the annulus (positive corresponding to ascending musical motion and counter-clockwise geometrical motion, negative descending and clockwise). This number can be taken to represent an element in the annulus's *fundamental group*, a topological object parameterizing the different voice leadings from Y to Y, and hence from X to Y.

We are now in a position to understand the phenomena in the previous section. Figure 10 uses annular space to model the two sequences in Figure 3: the first involves no cumulative angular motion and does not contain the annulus's center; the second makes a complete clockwise circle

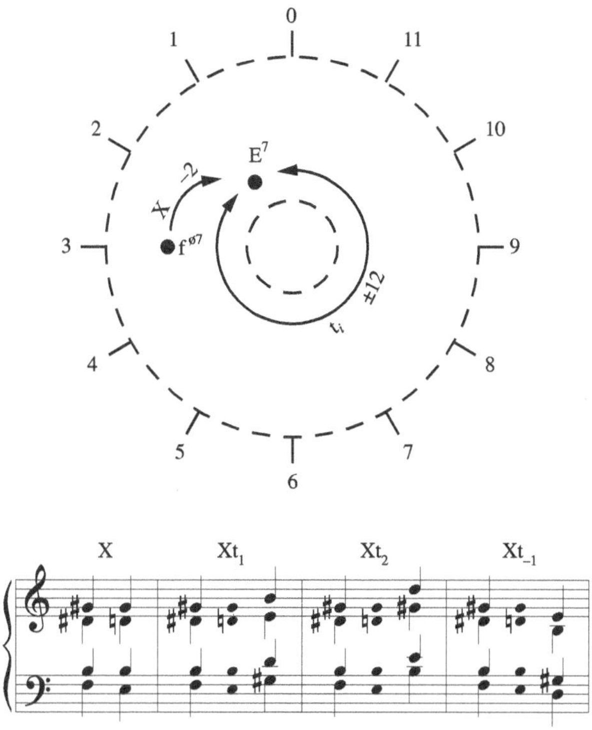

Figure 9. The voice leadings between two separate chords X and Y can be written as a fixed voice leading $X \to Y$ composed with some number of loops $Y \to Y$. In the annular space, the numbers represent sum of the real components in voice leadings X (which takes $f^{\varnothing 7}$ to E^7) and t_i (which are the transpositions up and down i steps along the chord E–G♯–B–D).

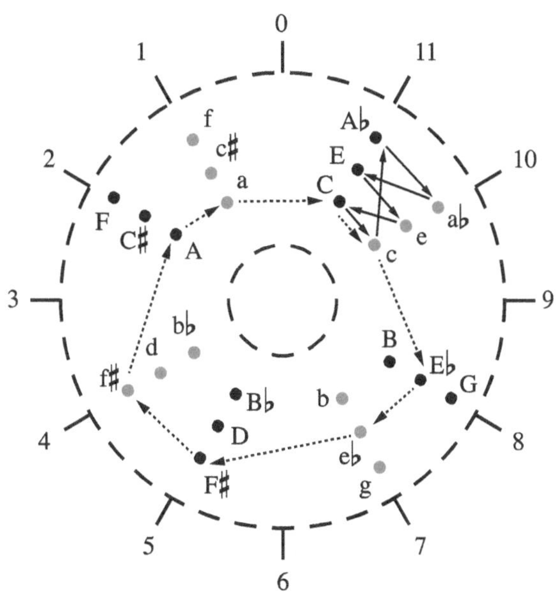

Figure 10. The paths in Figure 3 in annular space.

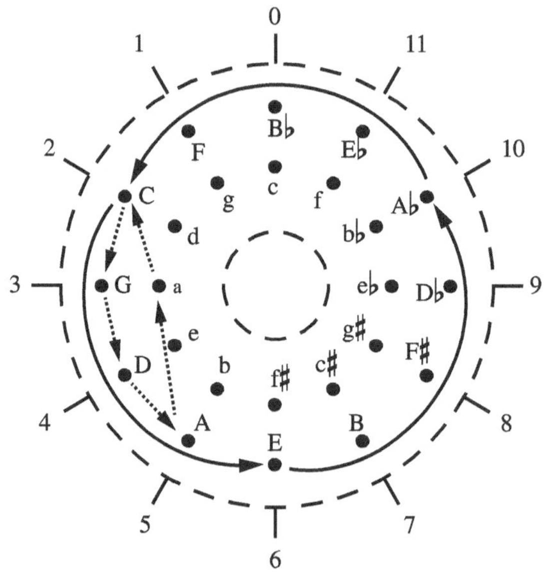

Figure 11. The paths in Figure 4 in annular space. Here I represent minor keys with harmonic-minor scales.

and moves each voice by one step along the chord.[15] Figures 11 and 12 repeat the demonstration for the "chart of the regions" and Lerdahl's diatonic torus respectively.[16] These graphs show that the annular topology can appear in our theoretical models *even if we do not explicitly put it there.* It captures a general and ineluctable feature of the space in which contrapuntal motion takes place.

Since the structure of the annular spaces is independent of the particular intervallic constitution of chord and scale, Figure 13 can be taken to represent the strongly crossing-free voice leadings connecting the transpositions of *any* three-note chord in *any* seven-note scale. To see this, let us place some form of our chord at the location marked C. It follows that the point marked B, 3/7 of the way clockwise around the circle, must represent its transposition down by scale step. (This is because the voice leading that lowers each note by scale step has paths whose real numbers sum to –3, and hence is represented by a clockwise path moving 3/7 of the way around the circle.) Continuing in this way, we see that diatonic trichords are evenly spaced around the circle, each a third below its nearest clockwise neighbor. It also follows that there is a transpositional voice-leading schema generating all the clockwise voice leadings on the figure; this is the motion that links one chord to its nearest clockwise neighbor, and whose retrograde links chords to their nearest counterclockwise neighbors. This *basic voice leading* always combines one-step transposition downward along the chord with two-step transposition upward along the scale – the former requiring seven total descending steps of motion and the latter six ascending steps, combining for one descending step in total.[17] We can obtain any path on the figure by repeatedly applying this schema or its retrograde.

---

[15] In Tymoczko (2010, 2011), I criticized the *Tonnetz* on the grounds that its distances do not fully match voice-leading distances; Cohn (2011a) answered by adding augmented triads to the *Tonnetz*. Here I raise a further issue that applies even to this modified *Tonnetz*, namely that it does not perspicuously reflect the underlying annular topology of the voice leadings it purports to model.

[16] Readers can verify these graphs using the multichord.html website described in footnote 5.

[17] Hook (2008) first considered basic voice leadings (called the "signature transformation") in the specific case of the familiar 7-in-12 diatonic scale; Hook (2011) later generalized this idea to arbitrary seven-note collections.

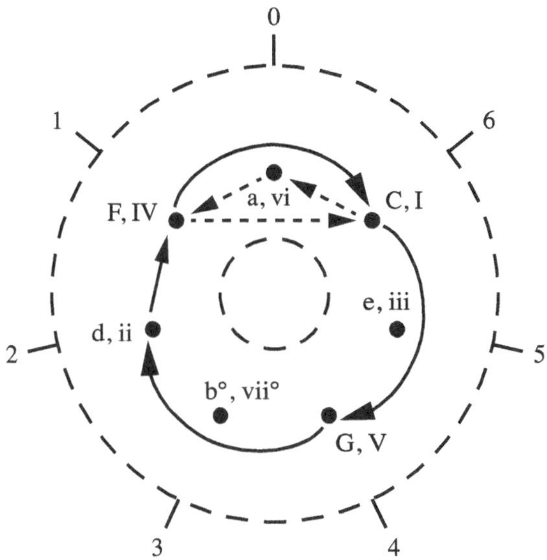

Figure 12. The paths in Figure 5 in annular space.

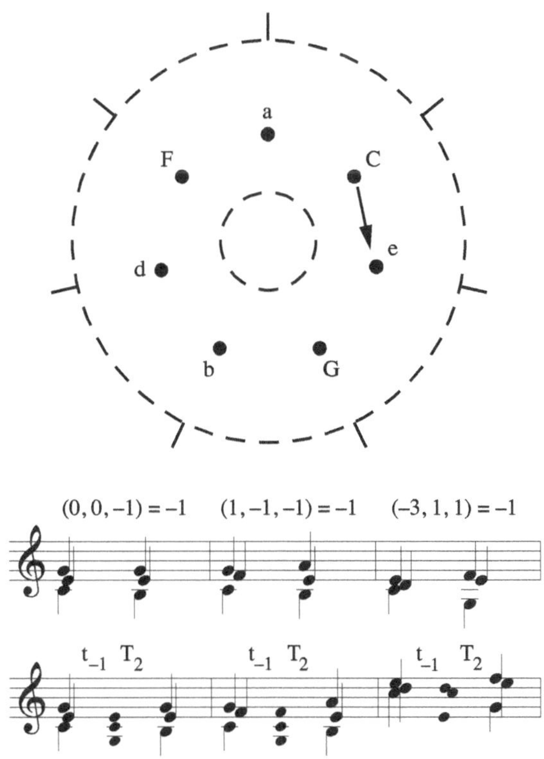

Figure 13. Three diatonic voice leadings that can be represented by the same graph. Lowercase t represents transposition along the chord; uppercase T represents transposition along the diatonic scale.

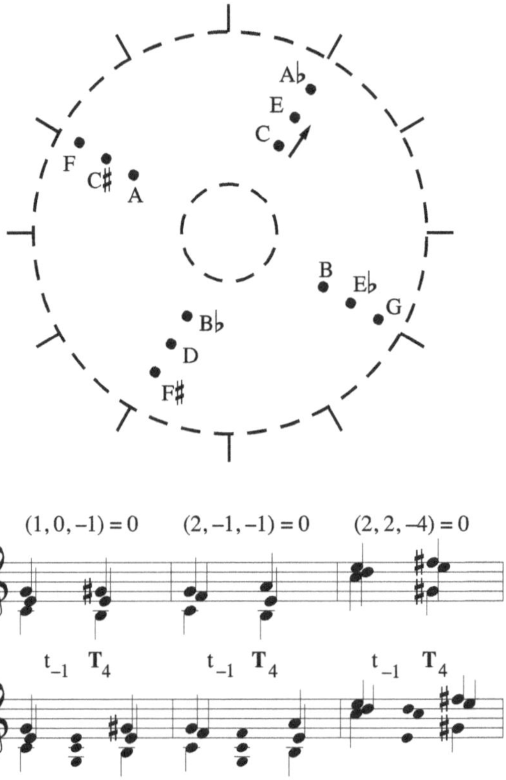

Figure 14. Three chromatic voice leadings that can be represented by the same graph. Lowercase t represents transposition along the chord while boldface **T** represents transposition along the chromatic scale.

Contrast this with the graph of any three-note chord in any twelve-note scale (Figure 14). Here we have three chords occupying the same angular position – from which it follows that there will be a transpositional voice-leading schema that, having been applied three times, returns each voice to its original position. Once again, this is true regardless of the structure of the chord and scale. This voice leading simultaneously transposes the chord by one third of an octave (e.g. up four semitones) while moving the voices along the chord by one third of an octave in the opposite direction (e.g. down a chordal step). (Since it involves two countervailing motions by one third of an octave, I call it a *diagonal action.*[18]) One of these involves an octave of ascending motion while the other involves an octave of descending motion, canceling out to produce a purely radial path. Once again this contrapuntal logic is independent of the chord's intervallic structure.

The annular representations suggest a number of compositional applications. First, as mentioned in the introduction, it gives us a notion of registral inversion general enough to be useful to twenty-first century musicians. Students are taught to construct inversions by moving the bottom note of a chord to the top, a procedure that works in the conventionalized language of classical tonality – where harmonies are always familiar, bass and melody are paramount, and the precise configuration of inner voices is often secondary. In modern contexts, when we are operating with a much wider range of chords and spacing may be important, the procedure breaks down (Figure 15). The loops in annular space suggest an alternative strategy of *transposing along the chord,* preserving its registral spacing as measured in chordal steps. I have found this to be a

[18] Mathematically, we have the cyclic group acting simultaneously as a transposition-along-the-scale and a transposition-along-the-chord (cf. Figure 22 below).

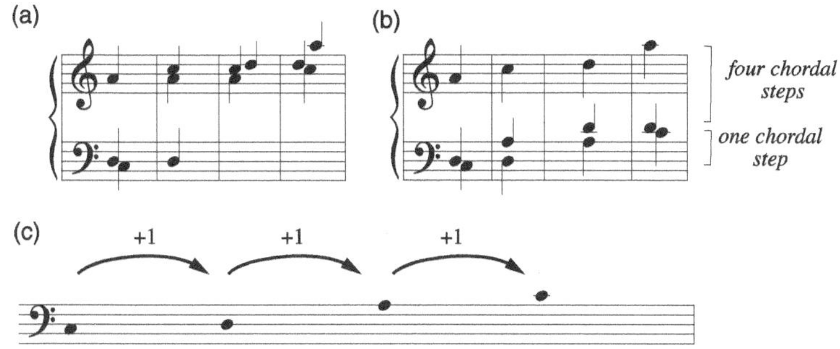

Figure 15. Two strategies of registral inversion. (*a*) Placing the bottom note of a chord at the top. (*b–c*) Moving each voice by one chordal step, preserving the chord's registral spacing when measured in steps along the chord.

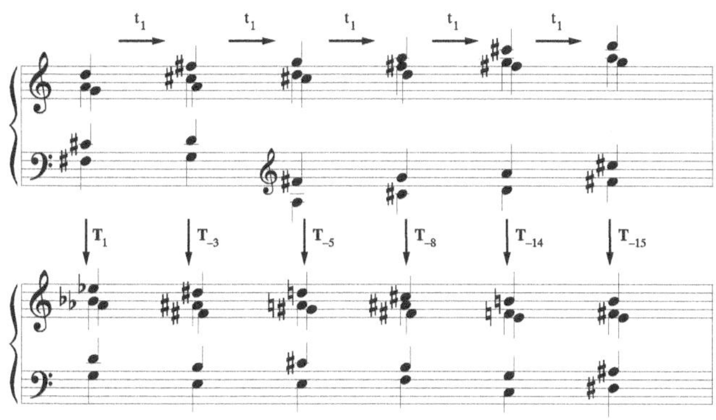

Figure 16. Generating a sequence of complex harmonies by combining motion along the chord with motion along the scale. The top system moves by one step along the initial chord. The bottom system transposes these harmonies chromatically to produce a progression of sonorities with the same spacing as measured in chordal steps.

useful technique for dealing with complex harmonies: when combined with scalar transposition, it generates a wealth of broadly similar sonorities from a single chord (Figure 16). Indeed, transposition along the chord and transposition along the scale are the two operations preserving transpositional set class and spacing as measured in chordal steps.

Annular space also allows us to transport musical procedures between musical domains: for instance, given a sequence of n-voice voice leadings linking the transpositions of some n-note chord in an $\boldsymbol{o}$-note scale, we can construct a structural analogue for any other n-note chord in any other $\boldsymbol{o}$-note scale, simply by moving the geometrical patterns from one annular space to another. Sometimes we can transport procedures between dissimilar domains – for instance, constructing broadly analogous passages that iterate two chords' respective "basic voice leadings." This in turn allows us to construct *hierarchical* passages that use similar contrapuntal techniques on multiple musical levels (Figure 17). Finally, if we take the repeated application of a transpositional voice-leading schema to represent a "generalized circle of fifths," we can use this structure to generate analogues of the melodic and harmonic minor collections, thus adapting traditional voice-leading relationships to arbitrary musical domains (Figure 18).[19] Understanding the compositional and perceptual significance of these ideas is a matter for future work; here I simply mention some potential compositional directions.

Analytically these tools show us how different genres make use of similar techniques. In a forthcoming book, I argue that there are many contexts in which we can find short clockwise

[19] When the initial schema is the basic voice leading for a maximally even chord, we can describe the intervallic structure of the generalized minor scales.

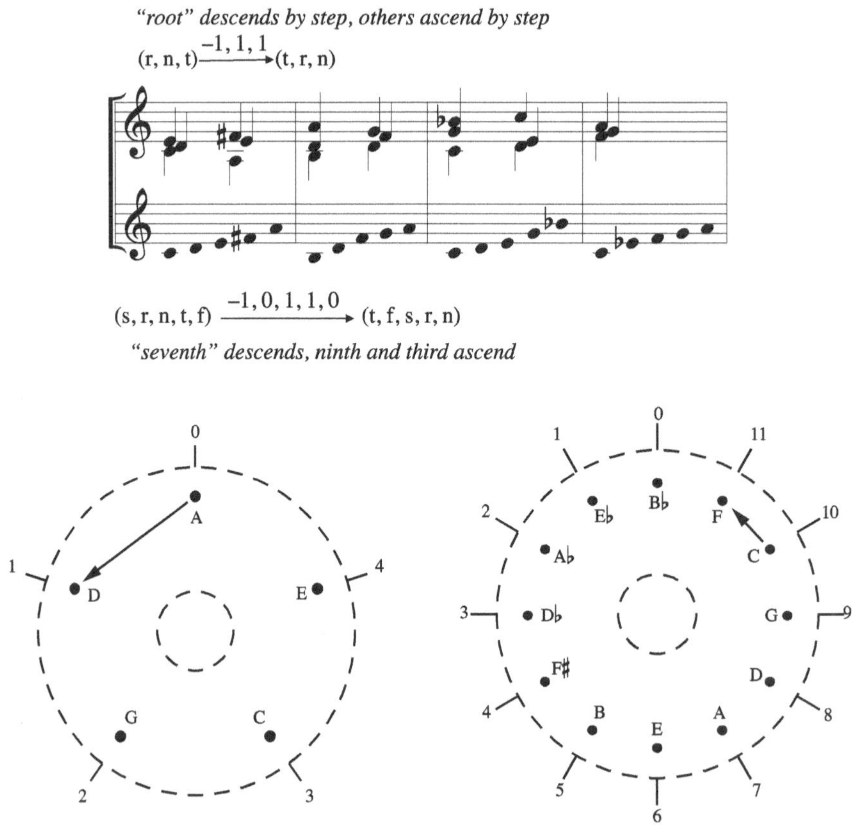

Figure 17. A hierarchical passage in which chords and scales each take short counterclockwise motions in their respective annular spaces. Here, *r*, *t*, *f*, *s*, and *n* stand for root, third, fifth, seventh, and ninth.

motions in the annular spaces, ranging from rock harmony to Renaissance counterpoint to classical modulation. (In this work I sometimes represent voice crossings by attaching loops as shown in Figure 19, using the spiral to indicate transposition; the result can be used to model *all* the two-voice voice leadings among the transpositions of any dyad in any seven-note scale.) Here geometrical music theory approaches the generality of Lewin's transformational theory, providing tools that apply to any chord and scale. What makes this possible is the realization that we do not need to represent voice-leading spaces in all their higher-dimensional glory; instead, we can use simpler and more abstract representations containing only the harmonies we happen to be interested in. My own experience is that this simplification greatly boosts analytical intuition, clarifying relationships that are much harder to see in the more complete geometry of chord space.

3. Topological models of transpositional set-class space

Musical geometry's core claim is that n-note chords live in the higher-dimensional analogue of a donut, with one dimension circular and the remaining n–1 dimensions forming a mirrored simplex. The annular spaces prioritize the circular dimension, using a single line segment to represent the rest. Another family of models prioritizes the simplicial dimensions, ignoring the circular dimension altogether. These are the three varieties of *set-class space,* the permutation

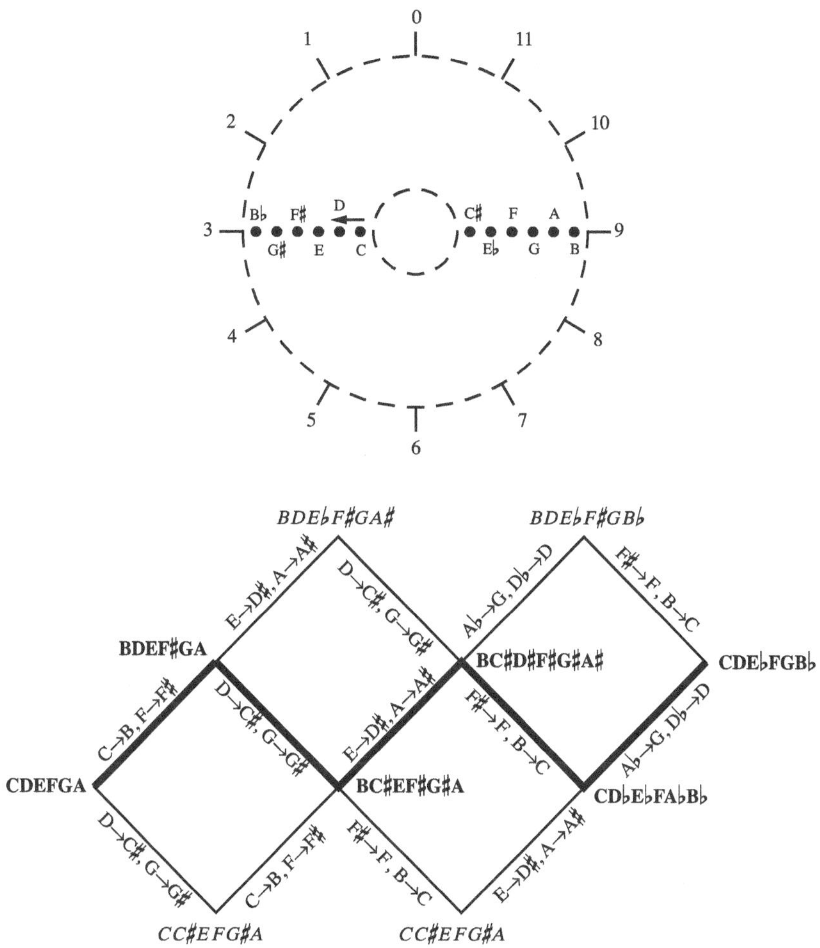

Figure 18. (top) Annular space for the diatonic hexachord. (bottom) Scrambling the voice leadings in any repeating voice-leading pattern produces analogues to the melodic minor collection. Here, the boldfaced chords represent a generalized "circle of fifths" that moves radially in annular space; by rearranging every pair of these voice leadings we obtain alternate pathways, generating hexatonic scales. See Tymoczko (2011, §3.11) for more.

region, transpositional set-class space, and inversional set-class space. Set theory and the annular spaces thus arise from complementary processes of abstraction, one deemphasizing voice exchanges, the other transposition.

In "Generalized Voice-Leading Spaces," Clifton Callender, Ian Quinn, and I described these n-note set-class spaces as $(n-1)$-dimensional simplexes, or generalized tetrahedra, with some of their points acting like mirrors or being glued together. (This description is mathematically awkward as the identifications form spaces that are no longer meaningfully "simplicial," but it has the advantage of providing a way to picture them.) Once again we associated voice leadings with the paths arising when each note makes a smooth glissando from its starting point to its destination.[20] In lower dimension these spaces can simply be drawn, allowing theorists to directly observe the relevant musical trajectories. In higher dimension the graphical strategy is unavailable, yet Callender, Quinn, and I provided no useful alternative. For this reason it would be fair

[20] To be more accurate, this earlier work associated voice leadings with "generalized line segments," the paths arising when each note glides linearly from its starting point to its destination, beginning and ending at the same time. This was to avoid defining the fundamental group at singular basepoints. See the appendix for more.

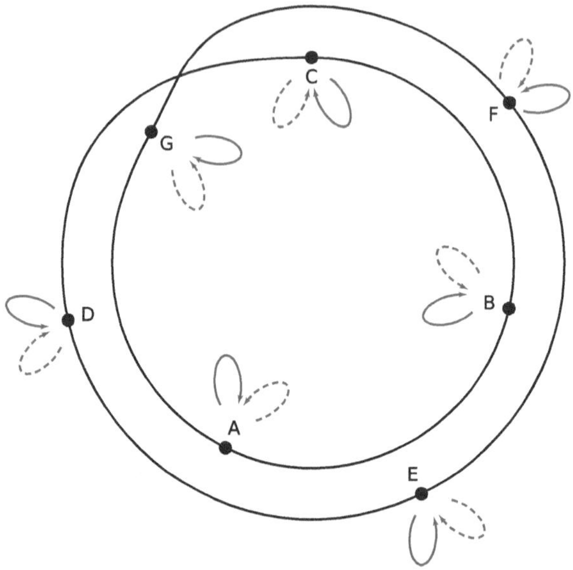

Figure 19. Annular space for dyads in a seven-note scale, with the loops representing two basic voice exchanges. For the diatonic third, these move notes by two and five contrary-motion steps respectively.

to say that our work stopped prematurely, couching our description of set-class space in terms that were too complex to be musically useful.[21]

In transpositional set-class space, points represent equivalence classes of chords related by transposition. Paths represent equivalence classes of *individually T-related* voice leadings – that is, voice leadings equivalent under the independent transposition of their chords (Figure 20). Since these are the set-class analogues of voice leadings, I will refer to them as "voice leadings between set classes" or just "voice leadings" when the context is clear. The group of n-voice voice leadings from a transpositional set class to itself can be decomposed into two familiar subgroups: the *voice exchanges,* generated by the n pairwise swaps of adjacent notes, with voices moving by the same distance in opposite directions, and the previous section's *transpositions along the chord.*[22] As before, the voice leadings between set classes X and Y can be represented by joining a default voice leading $X \to Y$ (usually taken to be a small perturbation, if one exists) to the complete collection of voice leadings $Y \to Y$, the fundamental group of set-class space. Thus we use the voice leadings from a set class to itself to understand the entire space.[23]

For an n-note chord or set class there are n fundamental transpositions-along-the-chord which I will label t_i, with the subscript corresponding to the number of steps. (I use an uppercase T for traditional transposition along the scale.) These compose additively for both chords and set classes, so that the result of following t_i with t_j is t_{i+j}. However, addition works differently in the two cases: with n-note chords, the n-fold application of t_1 moves each voice up by octave, which is not the same as t_0; in set-class space, the n-fold application of t_1 is exactly equal to t_0 since we factor out transposition (Figure 21).[24] Transposition along the chord thus determines an action of the cyclic group $\mathbf{C}_n$ in set-class space but not in chord space. (Somewhat surprisingly, this is the

[21] Again, identifications and singularities mean that these spaces can be topologically quite complex: for example, four-note transpositional set-class space is a cone over the real projective plane with singular base and tip (Callender, Quinn, and Tymoczko 2008).

[22] See the appendix and Tymoczko and Sivakumar (2018).

[23] My approach here is broadly reminiscent of Roeder (1984, 1987).

[24] For an n-note chord in an o-note scale, transposition by n chordal steps is equivalent to transposition by o scale steps, and this is reflected by the topological equivalence of the associated paths.

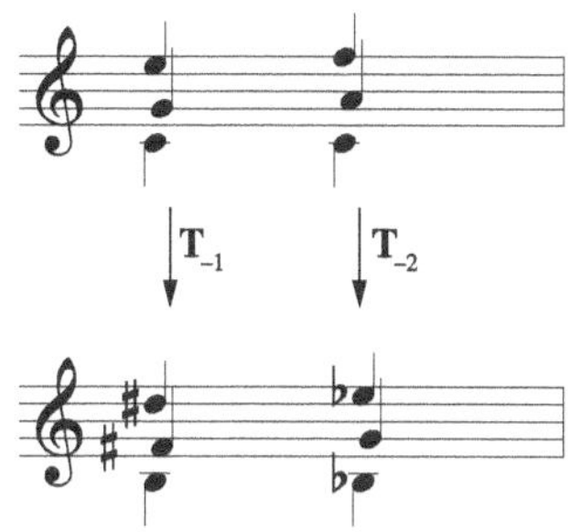

Figure 20. Individually T-related voice leadings, representing the same voice leading in transpositional set-class space. The chords in the top voice leading are transposed by different amounts to produce the bottom voice leading. Nevertheless, both begin with major chords, moving root by c, the third by $c + 1$, and fifth by $c + 2$. A single path in set-class space represents all such voice leadings, starting at any major chord and for every value of c.

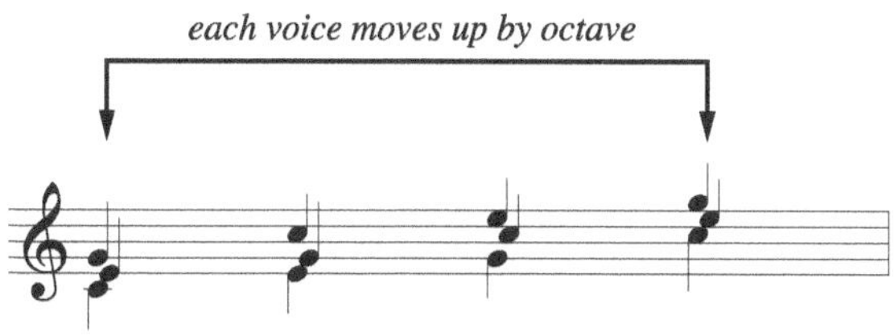

Figure 21. In trichordal set-class space, three ascending steps-along-the-chord are equal to the identity.

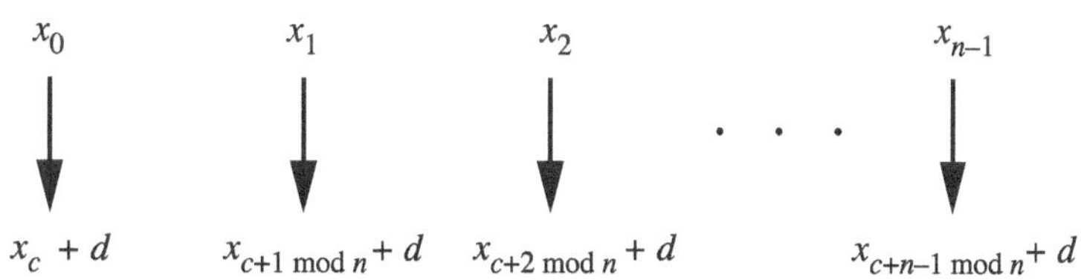

Figure 22. Let $(x_0, x_1, \ldots, x_{n-1})$ be a chord in non-descending pitch-class order spanning less than an octave. The combination of transposition along the chord and transposition along the scale moves the pitch class x_i to the pitch class $x_{i+c \bmod n} + d$, with addition acting on both pitch classes and voice labels. Voices move along paths $||x_{i+c}-x_i||_+ + d$, where $||x_{i+c}-x_i||_+$ is the *ascending c-step scalar interval* from x_i to x_{i+c}, which is $x_{i+c}-x_i$ if $i + c \geq i$ and $x_{i+c}-x_i + o$ otherwise. In set-class space we ignore d, while in the annular spaces we do not.

only difference between the two spaces' fundamental groups.) Thus subscripts in set-class space should always be understood modulo n: t_{-1} and t_4 are equivalent for pentachords. An interesting property of the transformations $t_i\mathrm{T}_j$, which combine transposition along both chord and scale, is that they apply addition to both pitch classes and voice labels (Figure 22).

An n-voice voice leading traces a path through an n-dimensional simplex with various identifications and mirror singularities. Because of these, the path might seem to disappear off one boundary to reappear on its identified boundary, or bounce off a singular point that acts like a mirror. The central claim of this section is that there is a principled association between boundaries and voice-leading transformations: in every transpositional set-class space, one boundary, the "permutation boundary," represents a voice exchange, swapping the set class's closest notes, while the remaining boundaries represent the nontrivial transpositions-along-the-chord. Geometrically, the permutation boundary acts like a mirror while the transpositional facets are identified pairwise, with the t_x and t_{-x} facets glued together. (In even dimension half of the $t_{n/2}$ facet is glued to the other half.) The claim that the boundaries are "associated with" a particular transformation has two meanings. First, that the transformation describes the structural changes occurring when any glissando encounters that boundary – for example, changes in the ordering of its voices by pitch-class, or the location of its smallest interval. Second, and somewhat more intuitively, that any path that starts at the interior of the simplex, moves to that boundary, and returns to its starting point (either reflecting off a mirror singularity or disappearing off the boundary to reappear on an identified face), articulates the associated transformation.

This leads to a graphical strategy in which we replace higher-dimensional "simplexes" with two-dimensional polygons (Figure 23), labeling each edge with its associated transformation – which is to say, the element of the fundamental group associated with loops passing only through that facet of set-class space. In higher-dimensional voice-leading space, each path can be associated with a unique voice leading between set classes; in the polygonal models, paths instead correspond to a broad class of topologically equivalent voice leadings (Figure 24). What is surprising is that there is only a small loss of analytical power resulting from this reduction to two dimensions. In large part this is because the boundaries are the spaces' most relevant features, musically and topologically.

Before diving into the details, it will be useful to consider a specific example. Figure 25 presents the opening progression of Wagner's *Tristan*. Callender, Quinn, and I associated this progression with a generalized line segment in our tetrahedral model of four-note set classes; in that model, this path originates at the point representing the half-diminished seventh, moves to the boundary at A where it disappears, reappearing at B to move down to C, "bounces off" the bottom edge to return to A, disappears again to reappear at B, and moves finally to the point representing the dominant-seventh chord. (This path, like most of the others in this paper, was determined computationally, with each voice gliding smoothly from its initial pitch to its destination over the same span of time.) To a musician this path is relatively uninformative. More useful is the knowledge that it first touches the t_2 boundary, then the voice-crossings boundary, and finally t_2 again. Figure 25c diagrams the path in polygonal space. We can think of the pattern of boundary interactions, t_2ct_2, either as recording the changing structure of the chord as the

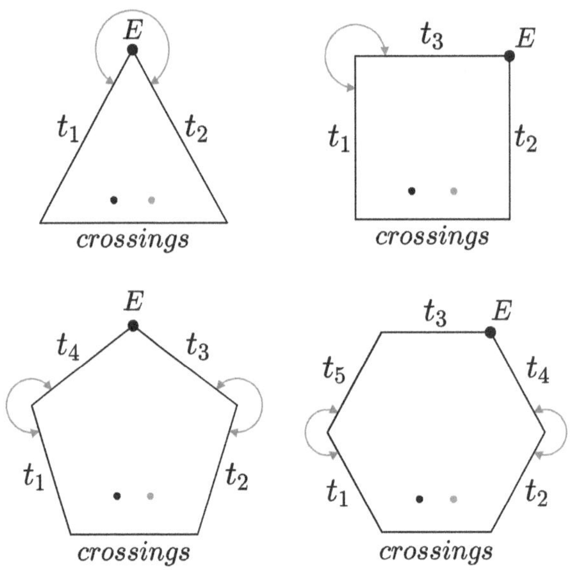

Figure 23. Polygons representing set-class spaces for three, four, five, and six notes. Dark dots represent a chord's normal form, light dots its inversion; arrows represent sides that are glued together. *E* is the completely even chord dividing the octave into *n* pieces.

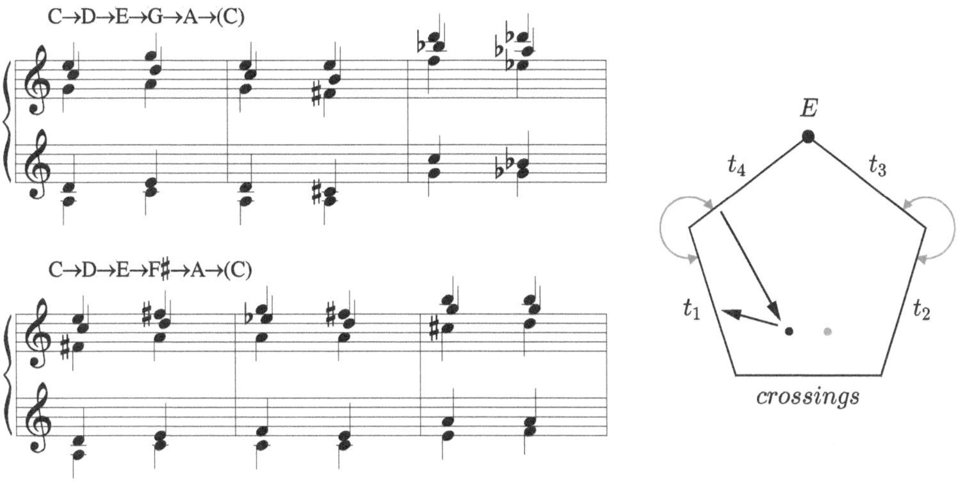

Figure 24. Pentachordal transpositional set-class space, with a path representing one-step transposition-along-the–chord. If we choose the dark point to represent the standard pentatonic scale, then this path will correspond to all of the voice leadings on the top system; if the point represents the dominant ninth chord, then the path corresponds to those on the bottom.

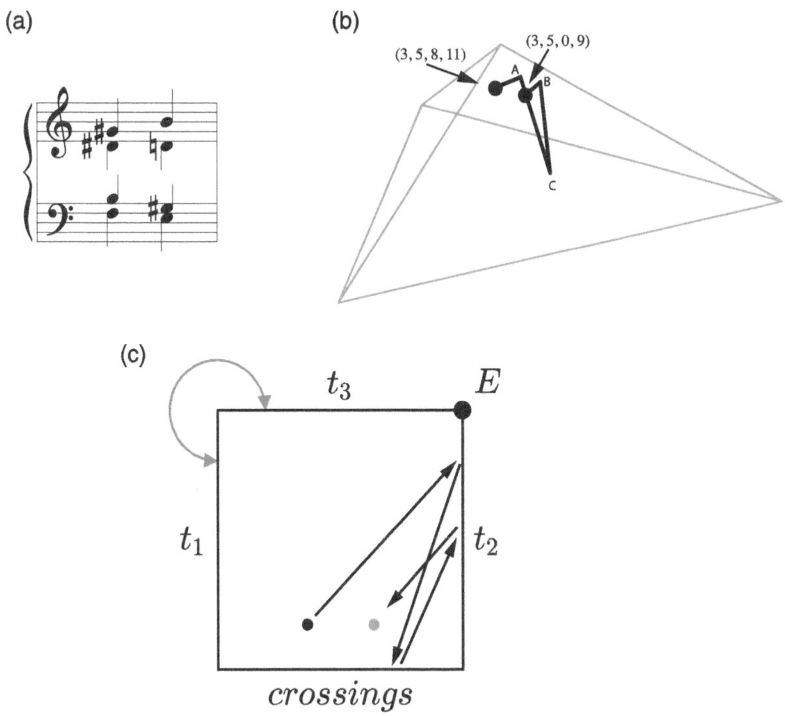

Figure 25. (a) The opening voice leading of *Tristan*, (b) the path it takes through tetrachordal transpositional set-class space, and (c) its appearance in the polygonal representation.

voices glide smoothly from start to finish, or more abstractly as transformations applied to the initial chord. I will consider each interpretation in turn.

First, the geometrically faithful approach. Figure 26 shows some of the harmonies that result from a smooth glissando taking the first chord to the second. For simplicity I transpose the

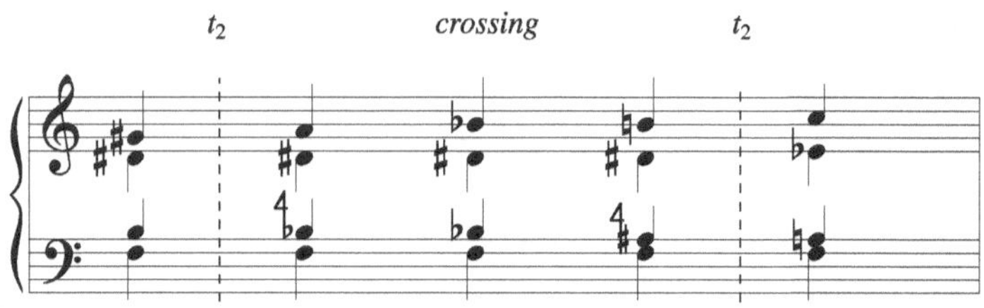

first chord in normal order							
	11	*10.66*	10.5	10	9.5	*9.33*	9
	8	*8.66*	9	10	11	*11.33*	0
	5	*5*	5	5	5	*5*	5
	3	*3*	3	3	3	*3*	3

Figure 26. The opening voice leading of *Tristan* with the second chord transposed up by semitone, and each voice making a smooth glissando from start to finish. The glissando touches the t_2 boundary when the moving voices are two semitones apart and the crossing boundary when they sound the same pitch class.

destination chord from E to F, since we are in set-class space where absolute transpositional level does not matter. In geometrical music theory, a chord's "normal form" is determined by the position of its smallest interval, as we will discuss shortly. We start with (3, 5, 8, 11), a half-diminished set class whose smallest interval lies between its first two notes; over the course of the voice leading, the G♯ glides up four semitones while the B glides down two. This path intersects with the t_2 boundary at the moment where the chord has two distinct "smallest intervals" (both of size 2), appearing between the first and final pairs in (3, 5, 8.66, 10.66); after interacting with this boundary, the sole smallest interval lies between the third and fourth notes. We can understand the t_2 label as representing the shifting position of the set class's smallest interval, from the first two voices up two chordal steps to the final two. The interaction with the voice-crossing boundary occurs when these two closest voices sound the same pitch class at (3, 5, 10, 10); after this point the third voice has moved above the fourth. (Here the voice-crossing boundary is associated with the fact that we need a different permutation to put the voices in ascending pitch-class order.) The second t_2 interaction occurs when the third voice is two semitones above the fourth, at (3, 5, 11.33, 9.33), with the chord again having two "smallest intervals" when ordered by pitch class. After that point, the smallest interval lies exclusively between the first two notes, as at the beginning. (Hence t_2 once again moves the smallest interval by two chordal steps.) From this point of view, the transformations record the general changes of the chord's structure – both its ordering in pitch class and the location of its smallest interval – that result when we represent the voice leading (3, 5, 8, 11) → (3, 5, 9, 12) as a glissando, sliding root to root, seventh to seventh, third to fifth, and fifth to third.[25]

Alternatively, and more intuitively, *we can interpret the geometrical pattern of boundary interactions as a sequence of abstract transformations combining to produce the total voice leading.* Figure 27 represents the transformational labels as voice leadings from the Tristan chord to itself, with the geometrical path now returning to the initial Tristan chord (the dark dot) after every boundary interaction, ending with a short path from the Tristan chord to its inversion. The path in Figure 27b is topologically equivalent to that in Figure 25c and represents the same total voice leading. We start with a t_2 that moves each voice up by two chordal steps, so that the seventh lies between soprano and tenor; these voices then cross by reflecting off the permutation boundary (which always exchanges the chord's closest notes along the shortest possible path); the music then reflects again off the t_2 boundary to return the seventh to the bass/alto pair before coming to rest at a different set class, here the inversion of our starting chord.[26] Once again we

[25] See the appendix for an explanation of the notation.

[26] In set-class space, any crossing of other pairs of voices requires a transposition along the chord that moves the chord's smallest interval so that it lies between the appropriate voices.

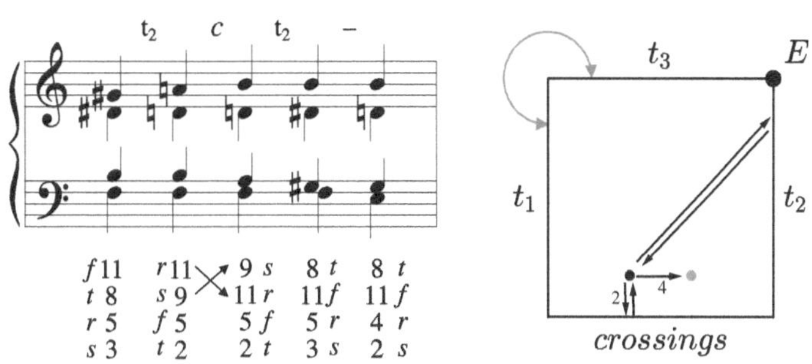

Figure 27. A transformational interpretation of the opening of *Tristan*, in which the boundary interactions are treated as voice leadings from the Tristan chord to itself. Here *r, t, f, s* stand for root, third, fifth, and seventh.

use the voice leadings $X \rightarrow X$ to understand the space, analyzing the Tristan voice leading as a series of voice leadings from the half-diminished chord to itself, coupled with a single "default" voice leading $X \rightarrow Y$.[27] The geometrical path – which was determined by a computer – thus gives us an *analysis* of the Tristan voice leading into a sequence of transformational "moves," here interpreted as voice leadings from the Tristan chord to itself (with one final move connecting the normal-form Tristan chord to the normal-form dominant-seventh).[28] We have, in other words, identified a connection between the geometrical and the algebraic, represented respectively by boundary interactions and contrapuntal transformations.

None of this is surprising: once we associate glissandi with voice leadings and construct the configuration space of set classes, then everything follows from the definition of the fundamental group. What is interesting, rather, is that familiar mathematics leads to simple, novel, and practical analytical tools. The next paragraphs will explain this connection in detail, describing the structure of transpositional set-class space, linking geometrical boundaries and contrapuntal transformations, and justifying the two-dimensional polygonal models. The main goal here is to show how paths through set-class space can be associated with sequences of transformations. Less mathematical readers may want to skip ahead to the conclusion of this section, circling back to the technical discussion as desired. (The polygonal models can be used even by those who do not follow the technical minutiae, treating them as convenient graphical representations of reasonably familiar voice-leading moves.) Mathematical readers may want to consult the appendix before proceeding.

* * *

A more detailed understanding begins with the notion of *geometrical normal form* as defined in Callender, Quinn, and Tymoczko (2008).[29] An n-tuple $(x_0, x_1, \ldots, x_{n-1})$ is in geometrical normal form if it meets the following criteria:

NF1. Its first element is 0;[30]

NF2. It is in ascending order, with the final element less than or equal to the octave $\boldsymbol{o}$;

NF3. Its smallest interval lies between its first two notes (including the "wraparound" interval $\boldsymbol{o}$–x_{n-1} among the chord's intervals).

[27] For convenience, I have chosen this final voice leading to be the unique crossing-free voice leading from Tristan to dominant seventh that preserves a chord's normal ordering, though we could in principle choose any voice leading between the two set classes. We could also apply the transformations t_2ct_2 to the dominant seventh instead.

[28] Readers can use the sc.html website to determine these paths.

[29] Thanks to Noam Elkies for helping Callender, Quinn, and myself understand musical geometry, and in particular the definition of geometrical normal form.

[30] NF1 provides the most useful set-class labels, at the cost of complicating the calculation of voice-leading distances; a less intuitive but geometrically more accurate alternative is to require that the set classes sum to a particular value, in which case we typically need negative numbers.

These equations determine a region of ordinary Euclidean space that can help us understand the more complicated geometry of set classes; the region represents set classes as n-tuples of real numbers in the range $0 \leq x \leq \boldsymbol{o}$, with 0 and $\boldsymbol{o}$ the same pitch class an octave apart.[31] The first important departure from traditional set theory is that we allow *multisets* or chords with multiple copies of the same pitch class. The second is that geometrical normal form is *non-unique:* a single set class can have multiple normal-form orderings, for example (0, 1, 2) and (0, 1, 11) in twelve-tone equal temperament. Where traditional theory mandates a process of tie-breaking to select one ordering as the unique normal-form representative, the geometrical alternative instead accepts that some set classes have multiple normal forms. We will see that this redundancy is actually a virtue of the geometrical approach.

Before considering set-class space in all its complexity, it is worth thinking about the simpler space that results from NF1–2. By elementary arithmetic, these define an $(n-1)$-dimensional simplex consisting of all n-tuples meeting the criteria, a Euclidean *fundamental domain* for a more complicated orbifold I call *the permutation region.*[32] Its vertices are the n combinations of 0 and $\boldsymbol{o}$ containing at least one 0 and sorted in ascending order; these each have a single nonzero interval of size $\boldsymbol{o}$, occurring in a different order position[33]

$$(0, 0, \ldots 0),\ (0, \ldots 0, \boldsymbol{o}),\ (0, \ldots 0, \boldsymbol{o}, \boldsymbol{o}), \ldots, (0, \boldsymbol{o}, \ldots, \boldsymbol{o}).$$

This simplex is the cross section of chord space, containing a separate point for each of the n modes of a given set class; thus the major chord is represented by (0, 4, 7), (0, 3, 8), and (0, 5, 9), all of which satisfy NF1–2.[34] Here we have a particularly clear example of the complementary relationship between the annular and set-class spaces, the former compressing the cross section into a single line segment, the latter ignoring the circular dimension in favor of the cross section itself. We can represent the cross section symbolically as an n-sided polygon whose sides exchange one pair of voices (Figure 28); together, the total collection of paths, starting and ending at the same point and reflecting off some series of boundaries, represents the subgroup of *voice exchanges.*[35] The space can be divided into n equivalent parts containing exactly one mode of each transpositional set class; each of these equivalent parts contains orderings whose smallest interval is in the same order position. (In analytical contexts we therefore have to make an arbitrary choice about which configuration of notes in a score represents "mode 1.") Despite its redundancy, this cross section usefully represents the various combinations of voice exchanges and transpositions-along-the-chord: for instance, analysts might prefer Figure 29 to the earlier depiction of the same progression in Figure 25c, as it eliminates all but one boundary interaction.[36] Here we see a trade-off between abstraction and analytical clarity, with the more abstract

---

[31] We will see that these octave-related pitches are needed to determine which boundaries are to be identified.

[32] In this context, a fundamental domain is a region of ordinary Euclidean space whose interior contains exactly one point for each point in the "interior" of the orbifold; it can be transformed into the orbifold by characterizing the singularities and indentified boundary points.

[33] The chord (0, 0, … , 0) has this interval in the "wraparound" position between its final note and the octave above its initial note.

[34] This simplex is closely related to what is called the *standard simplex* whose vertices are (1, 0, … , 0), (0, 1, 0, … , 0), … , (0, … , 0, 1). The entries in the standard simplex can be understood as the set-class's *step intervals* expressed as fractions of an octave (Regener 1974), whereas points in our simplex represent *pitch classes* (Tymoczko 2011, Appendix B); these differ only by a linear transformation. The standard simplex appears throughout applied mathematics, representing (for instance) the possible results of an n-candidate election.

[35] When voices cross, they also exchange their numbers: if voice 1 initially sounds C and voice 2 D, then voice 1 ends up sounding D and acquiring the label "voice 2" while voice 2 sounds C and acquires the label "voice 1." This relabeling ensures that the edge-labels remain accurate.

[36] In set-class space, there is only one voice-crossing boundary and it exchanges the closest pair of notes. Thus the musical function of the t_2 transformations is merely to cross the correct pair of voices. Mathematically, the pattern of boundary interactions analyzes the voice-crossing subgroup into combinations of transpositions-along-the-chord and a single crossing. The permutation region gives a more perspicuous representation of voice exchanges.

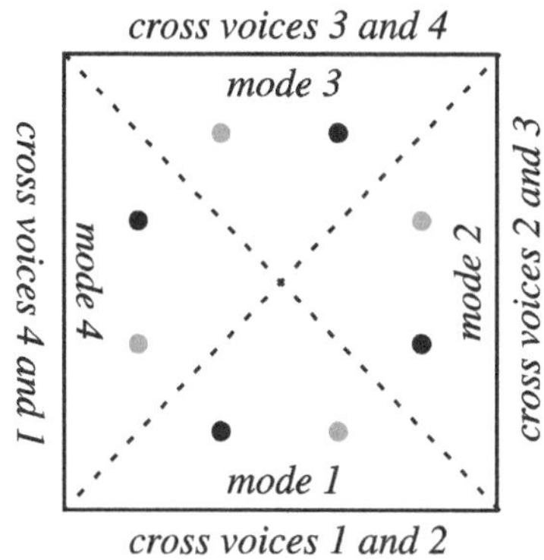

Figure 28. The permutation region represents the cross section of annular space; its n boundaries represent the n exchanges of pairwise voices. Each of its n regions contains a different mode of each set class.

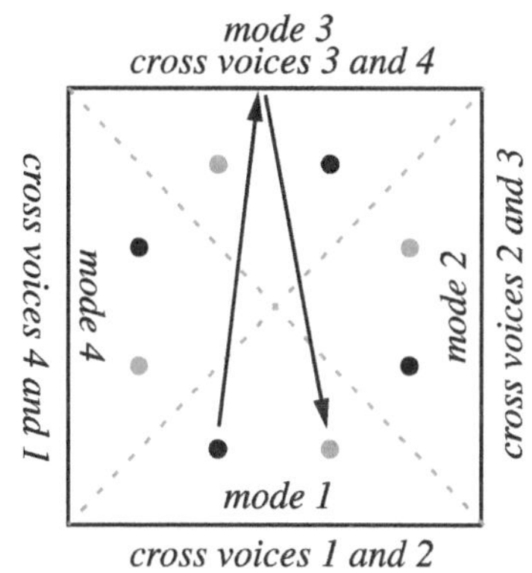

Figure 29. The first voice leading of Tristan plotted in the permutation region.

	2	3	4	5	6
v_1 (E)	0, 6	0, 4, 8	0, 3, 6, 9	0, 2.4, 4.8, 7.2, 9.6	0, 2, 4, 6, 8, 10
v_2	0, 0	0, 0, 0	0, 0, 0	0, 0, 0, 0, 0	0, 0, 0, 0, 0, 0
v_3		0, 0, 12	0, 0, 0, 12	0, 0, 0, 0, 12	0, 0, 0, 0, 0, 12
v_4			0, 0, 12, 12	0, 0, 0, 12, 12	0, 0, 0, 0, 12, 12
v_5				0, 0, 12, 12, 12	0, 0, 0, 12, 12, 12
v_6					0, 0, 12, 12, 12, 12

Figure 30. The vertices of transpositional set-class space for dimensions two through six.

set-class spaces representing familiar transformations in counterintuitive ways. We will return to this issue below.

To pass from the cross section to transpositional set-class space, we use NF3, which requires that the smallest interval be in the first position. This replaces the vertex $(0, \boldsymbol{o}, \ldots, \boldsymbol{o})$, whose largest interval is between its first two notes, with the perfectly even chord E dividing the octave into n precisely equal parts (Figure 30). A great deal of theoretical effort has gone into describing the internal structure of the resulting space, but there is an important sense in which this project is uninteresting: the interior of the simplex simply gives us the various ways to divide a single quantity $\boldsymbol{o}$ into n parts, with the stipulation that the first be no larger than any of the

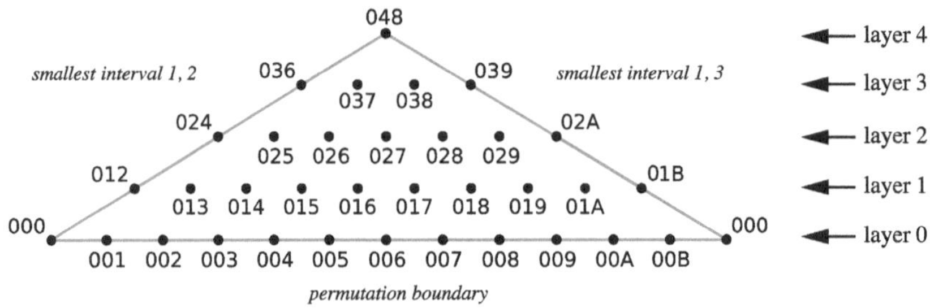

Figure 31. Transpositional set-class space has the structure of a cone.

others.[37] (This *banality of set-class space* is what allows us to represent the higher-dimensional simplex with a two-dimensional polygon.) There is nothing specifically musical about this structure, which could just as well represent groups of people of size ***o*** choosing among *n* mutually exclusive alternatives – say, the workers at an office splitting up to see movies at a multiplex. (This familiar geometry arises in music because a set class's *n* intervals divide the octave ***o***.) It is at the boundaries where we find the characteristically musical phenomena, "gluings" arising from octave and permutational equivalence.

Closer examination shows that the fundamental domain has the structure of a cone parameterized by the size of the set class's smallest interval (Callender, Quinn, and Tymoczko 2008). For example, the different layers of Figure 31 are essentially similar in structure, line segments whose endpoints are modes of the same set class. (These will therefore be glued together so that each layer forms a circle.) These layers are related by a set-theoretic analogue of the multiplicative (or *M*) transform that moves each chord linearly toward the perfectly even chord *E*, expanding intervals smaller than ***o****/n* while shrinking larger intervals, so that set classes get more and more even as we ascend toward the cone point.[38] We can use this conical structure by fixing the size of the smallest interval, drawing only one layer of some higher-dimensional space – as in Figure 32, which shows the semitonal layer of four-note set-class space.[39] Every four-note set class, other than the diminished-seventh *E*, is related by our multiplicative transform to exactly one point in this layer; consequently we can use it as a two-dimensional representation of four-voice voice leadings – quotienting out by our geometrical "*M* transform." As we will see, an analogous process gives rise to three-dimensional representations of five-note set classes, a four-dimensional space that is otherwise difficult to visualize.

Set classes in the interior of the space have no note-duplications and a unique smallest interval. One boundary, the *permutational facet*, contains set classes with more than one copy of some pitch class, with normal form $(0, 0, x_2, \ldots, x_{n-1})$. This is the base of the cone.[40] The remaining boundaries contain set classes with two or more instances of their smallest interval.[41] Since all

[37] See Roeder (1984, 1987), Cohn (2003), Callender (2004), Straus (2007) and Callender, Quinn, and Tymoczko (2008). The connection to the standard simplex is clearer when we switch from the note-based representation of set classes (e.g. 0, 4, 7 for the major triad) to Regener's step-interval representation (e.g. 4, 3, 5 for the major triad).

[38] This process is mathematically equivalent to the standard *M* transform when we use coordinates where the perfectly even chord is at the origin. In this coordinate system (1, 0, –1) is (C♯, E, G), (2, 0, –2) is (D, E, F♯), and (3, 0, –3) is (D♯, E, F), with the new coordinates related by the *M* transform and lying on the line segment from (C, C, C) to (C, E, G♯).

[39] The fundamental domain for each layer of *n*-note set-class space is similar to that of the $(n - 1)$-note permutation region described above. This is because the distance between the first two notes is fixed, leaving $n - 1$ voices to divide up the remaining portion of the octave.

[40] The term "facet" refers to the $(n - 1)$-dimensional boundary of an *n*-dimensional simplex; for a triangle it is an edge, for a tetrahedron a face, and so on.

[41] The presence of duplicate "smallest intervals" on the boundaries follows from NF3: the boundary is the place where one voice can no longer be raised (or lowered) and this is because doing so would form an interval smaller than the first with the next-highest (or next-lowest) note. This can also be seen by considering the vertices of each facet: the perfectly even set class contains the same interval at each order position, while the remaining vertices contain a unison at all but

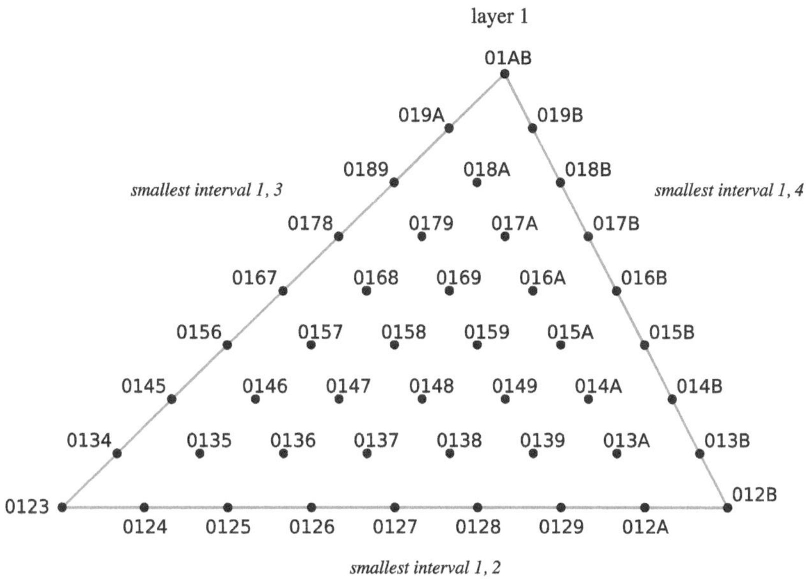

Figure 32. The semitonal layer of tetrachordal set-class space.

have one smallest interval in the first position (by NF3), they can be labeled by the position of their second smallest interval: one facet will contain chords whose second smallest interval is in the second position, another whose second smallest interval is in the third position, and so on. In Figures 31 and 32, the rightmost boundary contains chords whose second smallest interval is found in the "wraparound" position between its final note and the octave above its initial note.

These duplications determine how the simplex's facets are to be identified. Suppose any set class is in geometrical normal form and has its smallest intervals in the first two positions; then its one-step transposition-along-the-chord is also in geometrical normal form. For example, the one-step transposition of (0, x, $2x$, y, z), with x the smallest interval, is (0, x, y–x, z–x, $\boldsymbol{o}$–x), which is again in normal form.[42] The 1, 2 and 1, 5 facets must therefore be identified or "glued together," as they contain exactly the same set classes; the same is true for the 1, 3 and 1, 4 facets. Thus we see that $n - 1$ of transpositional set-class space's boundaries are identified pairwise, with the t_x and t_{-x} facets glued together. In even dimension, half of the facet representing $n/2$-step scalar transposition is identified with the other half, with chords like (0, 1, 5, 6) reappearing on the same face as (0, 1, 7, 8). This can be seen in Figure 33, where equivalent set classes are symmetrically arranged around the central vertical line containing transpositionally symmetrical chords.

If a facet has its smallest intervals in the $1, x + 1$ position, then its set classes will be kept in normal form by t_x, the voice leading that moves all notes up by x steps along the chord. A little thought will show that each facet acts like the *inverse* of this transposition, t_{-x}, in the following sense: a path that starts in the interior of the space and moves toward that boundary will reappear on its paired facet; if it then returns to its starting point, it will articulate the scalar transposition that preserves the normal ordering of the second, paired facet (Figure 34). This follows from the fact that line segments in the simplex represent voice leadings preserving normal ordering; to attach a pair of these, as in Figure 34, we need to transpose the second voice

one order position; a set consisting of the perfectly even set class and $n - 2$ of the remaining vertices will collectively have their smallest intervals at two common order positions. Linear interpolation will preserve this property.

[42] For a concrete example, transposing (0, 1, 2, 5, 7) by one step along the chord gives us another normal-form ordering, (1, 2, 5, 7, 12), or (0, 1, 4, 6, 11) when we start on 0. Here the one-semitone intervals lie in the first and "wraparound" positions (11, 12 with 12 being the octave transposition of the initial 0).

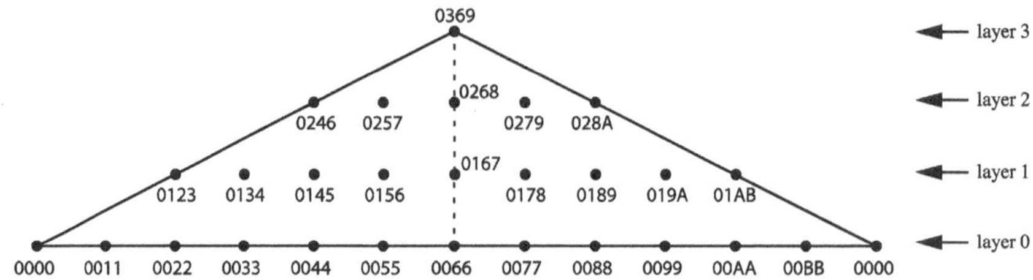

Figure 33. The 1, 3 face of tetrachordal transpositional set-class space. The central line is transpositionally symmetrical; each chord on the right side has a partner on the left.

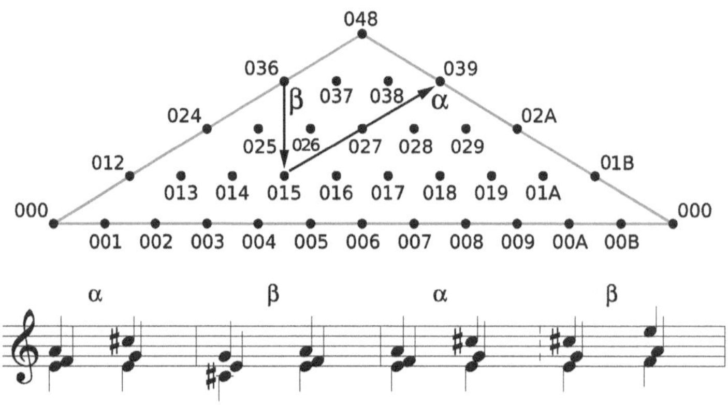

Figure 34. Two paths in trichord space, along with instances of the voice leadings they represent. We need to transpose β by one step upward along the chord to attach it to the endpoint of α.

leading along the chord so that it matches the first; the relevant transposition will necessarily be the one that keeps its chords in normal order.[43] (We can imagine the transposition occurs as we move off one boundary to reappear on the other.) Thus we can label each facet t_{-x} with x the transposition-along-the-chord that keeps its chords in normal order. A single loop will represent both ascending and descending x-step transposition-along-the-chord, with the musical direction-of-transposition corresponding to the geometrical direction-along-the-path. It follows that the strategy in Callender, Quinn, and Tymoczko (2008), of plotting voice leadings in the simplicial fundamental domain, determines a series of voice-leading transformations corresponding to the boundaries lying along the voice leading's path.

* * *

We have now completed our derivation of the polygonal graphs. Readers should be able to understand why the polygons in Figure 23 have one edge representing the voice exchanges (the base of the cone, containing chords with pitch-class duplications), and $n - 1$ edges representing the scalar transpositions, identified as appropriate. Motion within the polygon represents a voice leading from one normal form to another, changing the relative size of a chord's intervals while keeping the smallest interval in the first position. It should also be clear why a pattern of geometrical boundary-interactions (as in Figure 25b–c above) can be translated into a pattern of transformations (as in Figure 27), for the crossing of a transpositional boundary represents a shift

[43] The initial voice leading, moving from the starting point to the departure boundary, by definition ends with the chord in normal order.

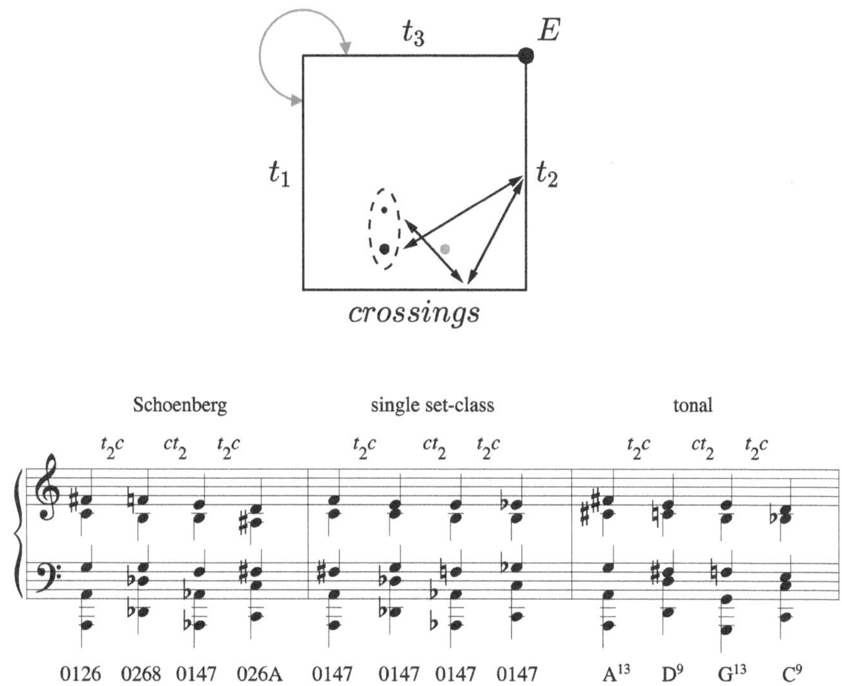

Figure 35. A single diagram in tetrachordal transpositional set-class space that can represent a four-chord progression from Schoenberg, a similar progression using just a single set class, or a descending-fifth series of extended dominants. In the first and third cases we conceive of our musical object as a region of set class space (the dotted oval); in the second, this region shrinks to a point, as in traditional set theory.

in the position of the set class's smallest interval; this is topologically equivalent to a transposition along the chord, coupled with a voice leading which remains within the interior of set-class space, preserving the normal ordering.

The main benefit of set-class space is to reduce a multitude of voice-leading possibilities to a much smaller number of basic templates. For an n-note chord in a $\boldsymbol{o}$-note scale, there are $\boldsymbol{n}\boldsymbol{o}$ distinct combinations of transposition-along-the-chord and transposition-along-the-scale (counting t_x and t_{x+n} as the same, since they differ only by an octave in each voice). By factoring out the transpositions-along-the-scale we reduce these to just n basic possibilities. If we want to consider distinct chord types we simply adjoin a single default voice leading $X \rightarrow Y$ to the total collection $Y \rightarrow Y$. In Tymoczko (2011) I have used this simplification for a variety of analytical and theoretical purposes, including (a) analyzing chromatic pieces that systematically explore the various voice-leading routes from one chord to another; (b) calculating the most efficient bijective voice leading from one chord to another; and (c) describing sequential passages with varying intervals of transposition. In the interest of space, I will not review these applications here.

A second use for transpositional set-class space, and for the polygonal models in particular, is in exploring voice exchanges. Since this raises interesting questions about the relative merits of the permutation region and transpositional set-class space, I will defer further discussion to §5.

A third use for these spaces is in modeling passages where chords are nearly but not exactly transpositionally related. Figure 35 shows the opening of the fifth song of Schoenberg's *Book of the Hanging Gardens* (Op. 15). Since each sonority belongs to a different set class, it might seem beyond the reach of conventional set theory; yet its three voice leadings each involve the exact same type of path in set-class space, interacting with both the permutation and t_2 boundaries. Figure 35 shows that we can conceive its four sonorities as small perturbations of a single

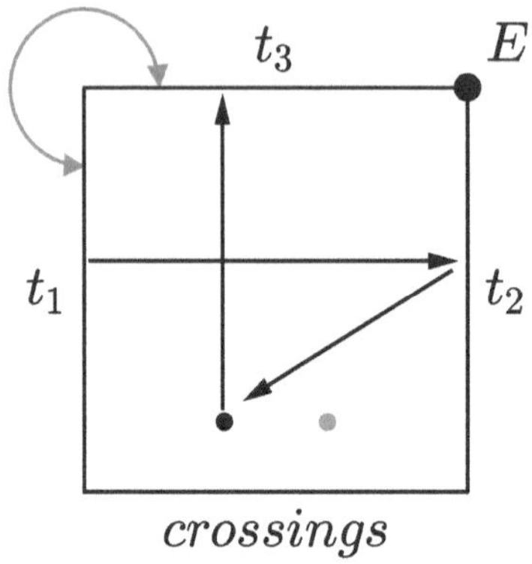

Figure 36. Linear interpolation (0, 1, 4, 6) → (1, 4, 6, 12) produces a generalized line segment that interacts with the t_3 and t_2 boundaries.

abstract structure – or better, as points in a region containing broadly similar set classes, each no more than two semitones away from 0147. (The graph encloses this region in an oval.[44]) The figure also shows a late-romantic descending-fifth sequence of extended dominants involving the very same pattern of boundary interactions.[45] Schoenberg's progression lies somewhere between these two models, at once distorting a tonal predecessor while pointing toward the more rigid structures of his later music. Without geometry we might be tempted to devalue such intuitions as "broadly similar harmonies" or "distortions of a familiar tonal schema" – notions that can seem defectively vague when compared to more precise music-theoretical language. But I would argue that such flexibility is fundamental to early atonality, requiring something like the analytical perspective we are exploring here.[46] One might say that geometry gives us a *harmonic* analogue to the theory of contour, defining flexible vertical structures analogous to contour theory's flexible melodic categories (Callender, Quinn, and Tymoczko 2008). And while this analysis could in principle use the complete tetrahedral set-class space, the two-dimensional representation is considerably more intuitive.

Three final points about the contrast between geometry and topology. In earlier work, I associated voice leadings with "generalized line segments," the specific paths that arise when each voice glides smoothly from its starting point to its destination. This can sometimes produce counterintuitive results: for example, the generalized line segment associated with the one-step ascending scalar transposition (0, 1, 4, 6) → (1, 4, 6, 12) interacts with two boundaries, decomposing t_1 as t_3t_2 (Figure 36). This because there is a point along the glissando where the smallest interval briefly lies between the second and third voices before moving to the "wraparound" position. In analytical contexts, we are typically more interested in the fact that a voice leading is an instance of t_1, not that linear interpolation happens to generate a somewhat counterintuitive path through set-class space for one set class but not another. Thus this paper associates voice leadings not with the specific paths resulting from linear interpolation but rather *homotopy classes* of paths all producing the same voice leading between their endpoints. Here topological abstraction may be preferable to geometrical particularity.

[44] Note that we cannot determine chord similarity by comparing normal orderings: sometimes, two set classes' normal forms can be relatively far apart, even while other modes of the chords are quite close. For instance, the normal form (0, 0.9, 6, 7, 8, 9) is farther from the normal form (0, 0.9, 2, 3, 6, 7) than it is from the non-normal (0, 1, 6, 6.9, 8, 9). My strategy can be adapted to such cases by choosing a region of space containing a transpositional boundary.

[45] See Stuckenschmidt (1965). This pattern can be considered a subset of a more complex five-voice structure has two voices sounding a fifth, two sounding a tritone, and an additional voice moving in contrary motion to sound chord roots.

[46] See Haimo (1996) and Callender, Quinn, and Tymoczko (2008) for a similar claim.

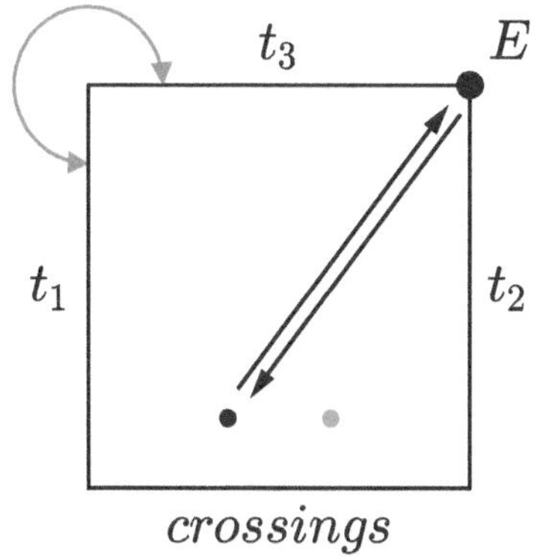

Figure 37. The effect of this path cannot be described topologically as it depends on the intervallic structure of the chord.

This same strategy is needed for voice leadings that pass through the highly singular tip of the set-class cone where the boundaries shrink to a single point. In the polygonal representations, as in the more complicated "simplicial" representations of Callender, Quinn, and Tymoczko (2008), voice leadings appear to "bounce off" this point to return to the interior of the space. However *the musical meaning of these paths depends on the structure of the chord.*[47] Thus the path in Figure 37 represents t_1 when we start with (0, 1, 3, 7), t_2 when we start with (0, 1, 3, 9), and t_3 when we start with (0, 1, 6, 8).[48] Intuitively, this is because the tip of conical set-class space can act as any of the transpositions-along-the-chord – that is any of the boundaries that shrink to this single point. Once again this is a reason to associate voice leadings with homotopy classes rather than specific paths: if a voice leading is associated with a particular path that passes through the tip of the set-class cone, then we cannot determine the contrapuntal transformation from its geometrical representation; if we instead associate voice-leading transformations with homotopy classes, then we can always find an unambiguous path representing a voice leading.

Finally, the polygonal models cannot faithfully model the intersections of the boundaries of the higher-dimensional spaces. This is because every pair of simplicial facets intersects, while only the adjacent sides of a polygon do. (For clarity, I arrange the polygonal graphs so the identified facets t_x and t_{-x} are adjacent.) Once again, this means that our polygonal diagrams cannot perspicuously model those specific paths that move through these highly singular boundary intersections; instead, we must represent them with topologically equivalent paths touching the relevant boundaries one at a time. As with voice leadings passing through the cone's tip, the algebraic language in §5 provides a useful alternative here.

4. Topological models of inversional set-class space

We now repeat the argument for inversional set-class space. Points here represent equivalence classes of tuples related by transposition, inversion, and permutation. Paths represent equivalence classes of *individually T-or-I-related* voice leadings – that is, voice leadings equivalent under the independent transposition of their chords, possibly combined with the inversion of

---

[47] Classifying these problematic paths is a topic for future work, requiring a more sophisticated approach to the singularities.

[48] Moving linearly toward E increases the size of any intervals smaller than $\mathbf{o}/n$ while decreasing the size of any larger intervals. After having moved through E, intervals continue to grow and shrink at the same rate, the largest interval in the initial chord becoming the smallest interval in the resultant. Exploring this continuous multiplicative transform is a matter for a future paper.

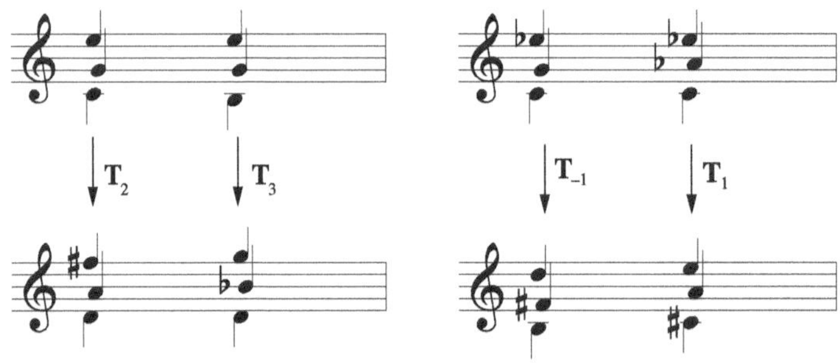

Figure 38. All of these voice leadings are represented by the same path in inversional set-class space. That path represents voice leadings that either start with a major triad, moving root by c and third and fifth by $c + 1$ semitones, or start with a minor triad, moving root and third by c–1 semitones and fifth by c semitones.

the entire voice leading (Figure 38). As before, the resulting space is a conical orbifold which can be described, loosely, as an $(n - 1)$-dimensional simplex with various singularities and identifications. The base of the simplex contains sets with note duplications and represents voice exchanges; the remaining boundaries now represent a host of dualistic transformations including *transpositional equivalence classes of bijective strongly crossing-free voice leadings between a chord and its inversion.* These are simultaneously the dualistic or inversional analogues of the transpositions-along-the-chord and also generalizations of the basic transformations of Cohn's neo-Riemannian theory. Understanding these transformations is the main obstacle to comprehending inversional set-class space.

Music theorists use the term "inversion" in a confusing variety of ways, even when we bracket the notion of registral inversion ("root position," "first inversion," etc.) to focus on inversion in pitch or pitch-class space. Theorists initially defined inversion as a global symmetry that turns musical space upside down, sending note x in voice A to its inversional partner $c - x$, once again in voice A; the parameter c, sometimes called the *index number,* was initially conceived as a fixed constant not depending on the structure of the chord. Later theorists defined *contextual inversions* in which the index number varies with the notes it is applied to; inversion in this sense is less a symmetry of musical space than a localized musical move, something you can *do* to a chord, as David Lewin put it. For instance, following Fiore and Satyendra (2005) we could define a contextual inversion that transforms an ordered pitch set $(x_0, x_1, \ldots, x_{n-1})$ into $(x_0 + x_1 - x_0, x_0 + x_1 - x_1, \ldots, x_0 + x_1 - x_{n-1})$. Here the index number $c = x_0 + x_1$ is "contextual" because it depends on the input ordering. (For exactly this reason, contextual inversion is not a global symmetry of musical space.) If we interpret order positions as voices then this contextual inversion again has the property of mapping note x in voice A to $c - x$ in voice A. Alternatively, we can apply contextual inversions to unordered chords by focusing on a smaller collection of sonorities and defining the inversions in structural terms – for instance, "the inversion that preserves the perfect fifth of a major or minor triad." Here again, the most straightforward interpretation is of an inversion operating individually on pitches or pitch classes, with the fixed point determined by the collection itself.

Generalized neo-Riemannian transformations, or bijective and strongly crossing-free voice leadings from a set class to its inversion, are neither traditional nor contextual inversions. Instead, they are transformations with the interesting property of acting analogously on pitch classes and voices, sending note x in voice A to note $c - x$ in voice $d - A$, where $d - A$ is a separate inversion applying to voice labels (Figure 39). In this respect they are similar to the previous section's transpositions-along-the-chord, which apply addition to both pitch classes and order numbers (c.f. the earlier Figure 22). This simultaneous action on both pitch classes and voice numbers is the crucial feature of Cohn's version of neo-Riemannian theory, encoded in

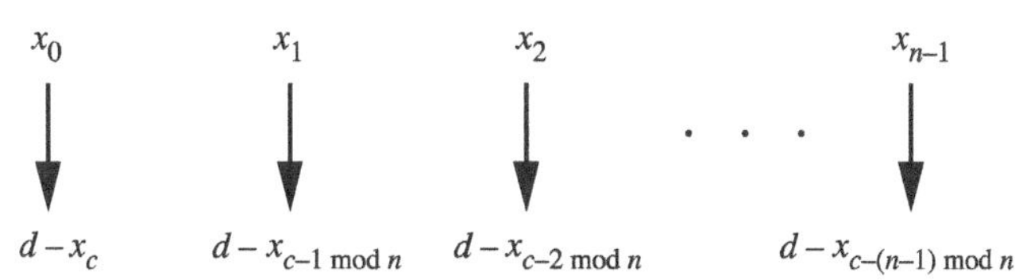

Figure 39. Let $(x_0, x_1, \ldots, x_{n-1})$ be a chord in non-descending pitch-class order spanning less than an octave. A strongly crossing-free voice leading from a chord to its inversion moves the pitch class x_i to the pitch class $d{-}x_{c-i \bmod n}$, with subtraction acting on both pitch classes and voice labels. Voices move along paths $d{-}x_c{-}x_i + ||x_c{-}x_{c-i \bmod n}||_+$, with $||x||_+$ defined as in Figure 22 .

the intuitive operation of a "triangle flip" (§1).[49] There is a close connection between this double action and voice-leading *efficiency,* as the operation on voices is capable of counteracting or undoing the operation on pitches. This is the conceptual link between Cohnian triangle-flips and the phenomenon of "voice leading parsimony" (Tymoczko 2008).

These generalized neo-Riemannian voice leadings will always preserve the distance between at least one pair of voices. When we are dealing with specific chords, we can define these transformations by choosing any two notes to be common tones: typically, there is a unique bijective and strongly crossing-free voice leading connecting the chord to its inversion and fixing those notes.[50] When we apply this voice leading to a specific registral configuration, we hold the chosen pitches constant, *preserving the chord's spacing when measured in chordal steps* – exactly as with the generalized "registral inversions" of Figure 15 (Figure 40). I have found this to be a fruitful compositional technique, for where traditional music theory tells us only how to invert chords in pitch- or pitch-class space, the neo-Riemannian approach generates a package of voicings that are broadly similar insofar as they share the same abstract spacing (Figure 41). By combining these neo-Riemannian voice leadings with transposition along the scale, we obtain all the strongly crossing-free voice leadings between the transpositions and inversions of a set class; in other words, exactly the voice leadings represented by the annular spaces of §2. Readers are encouraged to hone their intuitions by making use of the websites I have constructed.[51]

In set-class space, we need to consider equivalence classes containing all the voice leadings individually T-or-I-related to these. For want of a better term, I will call them "neo-Riemannian voice leadings in set-class space" or just "neo-Riemannian voice leadings" when the context is clear. We can label them relative to a chord's normal ordering: define $\mathfrak{i}_0$, as applied to normally-ordered S as the strongly crossing-free voice leading preserving the set class's smallest interval, sending its first note to the inversion of its second, its second to the inversion of its first, and so on (i.e. Figure 39 with $c = 1$ and d unspecified since we are in set-class space).[52] We define $\mathfrak{i}_i$, as applied to normally ordered S, as $t_i\mathfrak{i}_0$, combining $\mathfrak{i}_0$ with i-step transposition along the chord.[53] Since we are operating in inversional set-class space, we define these operations dualistically, so that $\mathfrak{i}_a(\mathbf{I}(S))$ is equal to $\mathbf{I}(\mathfrak{i}_a(S))$, with $\mathbf{I}$ representing traditional inversion (the global symmetry); this stipulation, that $\mathfrak{f}(\mathbf{I}(S))$ be equivalent to $\mathbf{I}(\mathfrak{f}(S))$ for some transformation $\mathfrak{f}$, is the defining feature of "dualism" as I use the term. (I will use a fraktur font for dualistic transformations.)

[49] Readers can verify that the three Cohnian triangle flips on the *Tonnetz* send note x in voice A to note $c - x$ in voice $d - A$.

[50] Exceptions can occur when the chosen notes are *o*/*x* scale steps and *n*/*x* chordal steps apart, in which case we can specify the inversions using the notation described momentarily.

[51] The nr.html website allows readers to explore the combination of neo-Riemannian voice leadings and transpositions, the tonnetz.html website represents neo-Reimannian inversions as polygon flips, the multichord.html website represents them in annular space, and the sc.html website displays neo-Riemannian voice leadings in set-class space.

[52] This definition is ambiguous for chords with multiple normal-form orderings; to label the contextual inversions in that case we need to arbitrarily choose an ordering. I will ignore this complication in what follows.

[53] This strategy is similar to Fiore and Satyendra (2005), only using a chord's normal ordering rather than an arbitrary ordering.

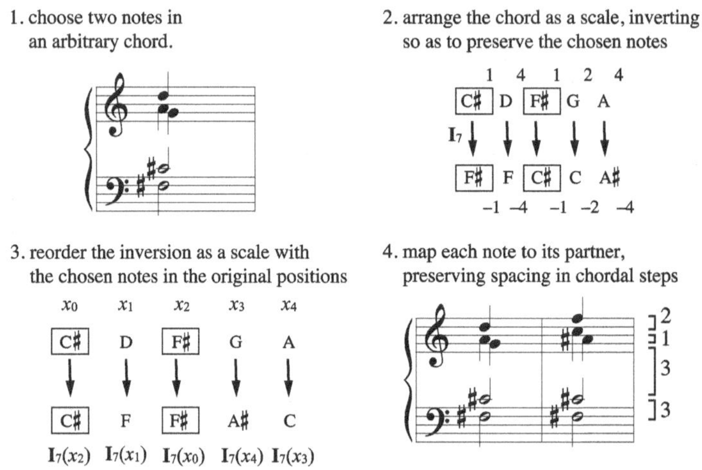

Figure 40. A recipe for constructing a generalized neo-Riemannian transformation.

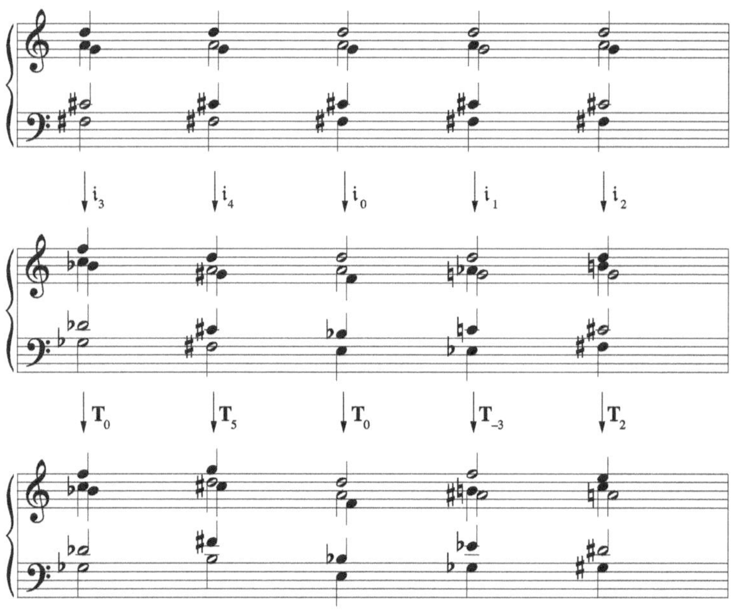

Figure 41. Combining neo-Riemannian voice leadings with chromatic transposition to generate a wealth of related sonorities from a single starting point. The inversional labels i_i will be explained shortly.

Since the inversion of an ascending a-step transposition-along-a-chord is a *descending* a-step transposition-along-its-inversion, this means that $i_a(\mathbf{I}(S))$ is $t_{-a}i_0$; it also implies that i_a is an involution, with $i_a i_a$ equal to the identity.[54] It follows that $i_0, i_1, i_2, \ldots$ map the first note x_0 of the normally ordered S to the inversion of its second, first, last, . . . , and third notes; thus each note x_i in the normal ordering is sent to $\mathbf{I}(x_{1-a-i})$, the pitch-class inversion of note x_{1-a-i}. For inversions, things work backwards: when applied to $\mathbf{I}(S)$, i_a maps the first note x_0 of its transpositional normal ordering to the inversion of its second, third, fourth, . . . , last and first notes, sending

[54] That i_0 is an involution, and that it commutes with the nondualistic transpositions-along-the-chord t_a, both follow from direct inspection; this proves that $i_a i_a$, or $t_a i_0 t_{-a} i_0$, is the identity.

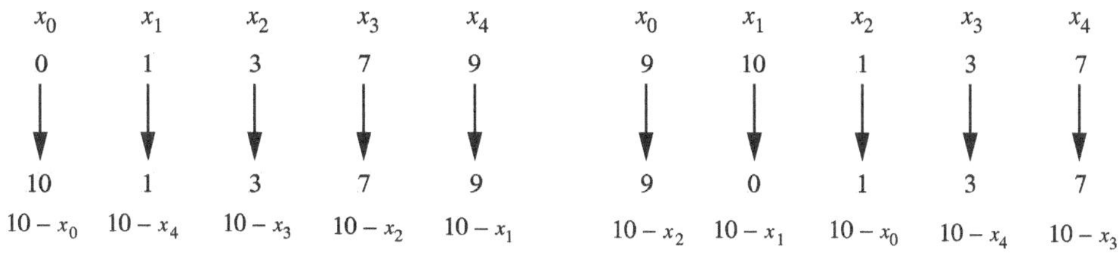

Figure 42. The operation $\mathfrak{i}_1$ applied to a normal-form chord (left), where it moves x_0 to x_0, and its inversion (right), where it moves x_0 to x_2. Each voice leading is the other's inverse. Subscripts are relative to each chord's transpositional normal form.

note x_i in the transpositional normal ordering to $\mathbf{I}(x_{1+a-i})$ (Figure 42). Under these definitions the neo-Riemannian L, R, and P voice leadings are instances of $\mathfrak{i}_0$, $\mathfrak{i}_1$, and $\mathfrak{i}_2$ respectively, whose set-class-space analogues we can write as $\mathfrak{l}$, $\mathfrak{r}$, and $\mathfrak{p}$.[55]

Where the n transpositions-along-the-chord combine like the cyclic group $\mathbf{C}_n$, the $2n$ neo-Riemannian set-class voice leadings combine like the dihedral group $\mathbf{D}_{2n}$, with an even number of inversions producing the "dualistic" transposition-along-the-chord $\mathfrak{t}_x$, ascending for the normal form and descending for its inversion. (Algebraically, $\mathfrak{i}_x\mathfrak{i}_y$ is equal to $\mathfrak{t}_{x-y}$ while $\mathfrak{i}_x\mathfrak{i}_y\mathfrak{i}_z$ is equal to $\mathfrak{i}_{x-y+z}$, or $\mathfrak{t}_{x-y}\mathfrak{i}_z$.) These somewhat counterintuitive relationships can be understood with the symmetries of a regular polygon: cut an n-sided polygon out of paper and label the vertices on one side with an n-note chord's notes in clockwise ascending order; on the other side of the paper, label the vertices with the inversion's notes in ascending clockwise order, with the two chords' smallest intervals sharing an edge. Voices correspond to the vertices' spatial position and neo-Riemannian set-class voice leadings to outline-preserving flips. Figure 43 demonstrates using boldface and italics for the two sets of labels. Two distinct flips form a rotation, sending the two sets of labels in the same spatial direction but opposite musical directions. (Transpositional set-class space contains only these rotations, recognizing no similarity between the chords on opposite sides of the polygon.) In odd dimension every flip preserves an edge while in even dimension only the even-numbered inversions preserve edges; odd-numbered inversions instead preserve two vertices and the distances between two pairs of notes that are adjacent-but-for-one-note. These flips are the larger-cardinality, set-class generalizations of Cohn's "triangle flip" operation.[56]

We will associate these dualistic transformations with the "boundaries" of inversional set-class space, leading to models like those in Figure 44. Here every inversion other than $\mathfrak{i}_1$ is associated with a unique facet of the space. A geometrical path through the space thus analyzes a voice leading into a series of neo-Riemannian set-class voice leadings just as a geometrical path through transpositional set-class space analyzes the voice leading into a series of transpositions-along-the-chord (plus voice exchanges, in both cases). Hence the neo-Riemannian set-class voice leadings are *the voice leadings that remain when we quotient out by transpositions, inversions, and voice exchanges,* much as transpositions-along-the-chord are *the voice leadings that remain when we quotient out by transposition and voice exchanges* (described as *R* in the introduction).[57]

[55] The triad's geometrical normal form is 037. L is $\mathfrak{i}_0$ or (0, 3, 7) → (0, 3, 8), moving 0 to $I_3(3)$, R is $\mathfrak{i}_1$ or (0, 3, 7) → (10, 3, 7), moving 0 to $I_{10}(0)$, and P is $\mathfrak{i}_2$ or (0, 3, 7) → (0, 4, 7), moving 0 to $I_7(7)$.

[56] Cohn's "triangle flips" associate voices with vertices and pitch classes with spatial positions. This approach is hard to generalize, first, because only the three-note *Tonnetz* tiles the plane, and second, because only in the three-note case is the number of edges equal to the number of pairs of notes. The polygonal models in this paper instead associate spatial positions with voices and vertices with pitch classes. The tonnetz.html website in footnote 5 realizes a general chordal *Tonnetz*. See also Tymoczko (2012).

[57] Around 2002, Callender and I puzzled over the relation between neo-Riemannian theory, his trichordal set-class space (Callender 2004), and my own scalar and interscalar interval matrices (Tymoczko 2008). This paper's answer is

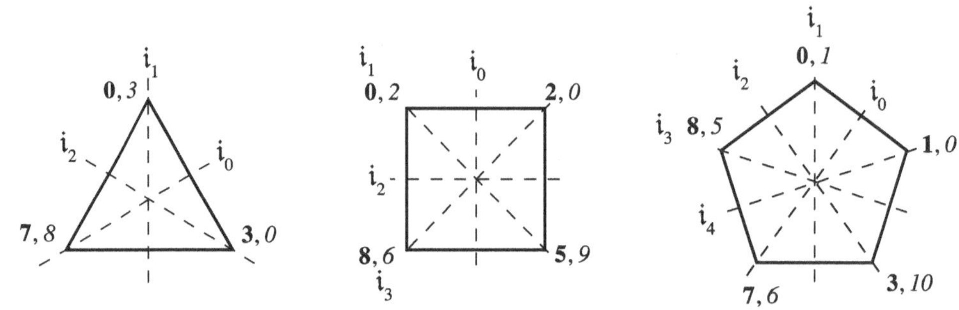

Figure 43. Polygonal models of the neo-Riemannian set-class voice leadings for the major/minor triad (left), the half-diminished/dominant seventh (center), and the 01378 pentachord in Figure 41 (right). Bold labels are on the front of the polygon, italics on the reverse side. The neo-Riemannian set-class voice leadings are flips around different axes; two such flips form a rotation that transposes the two sides in opposite directions along the chord.

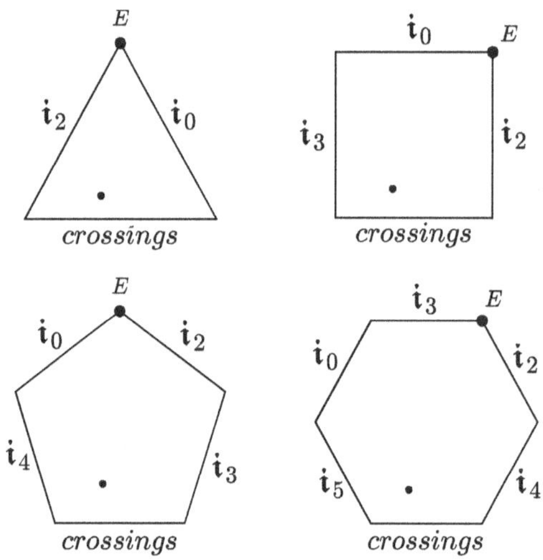

Figure 44. Schematic models of inversional set-class space, for trichords, tetrachords, pentachords, and hexachords. These graphs involve slight simplifications that will be described shortly.

Musically the generalized neo-Riemannian voice leadings are important because they preserve the distance between two voices. Such voice leadings often appear when composers move those voices in parallel, the remainder shifting so as to create strongly crossing-free voice leadings among inversionally related sonorities. Each of the passages in Figure 45 is of this form, repeatedly applying the same voice leading in set-class space.[58] Geometrically, this means that they each repeatedly move along a single line segment. (Collectively, the passages are related in a more general sense, making similar use of the space's different boundaries.) From a traditional standpoint, we can think of these voice leadings as repeatedly combining a familiar neo-Riemannian operation (either L, P, or R) with additional transpositions. Geometrically the i_0 and i_2 voice leadings are represented by paths that reflect off nearby edges before returning to

---

that the latter (in particular, the interscalar interval matrices relating inversionally related chords) depict the crossing-free subgroup of the homotopy group of Callender's trichordal set-class space, which in turn lifts to the Cohnian *Tonnetz*.

[58] These sorts of passages answer a delicate question in neo-Riemannian theory, namely "when should we assert the presence of a genuinely dualistic transformation like L, P, and R, rather than simply an efficient voice leading between inversionally related chords?" A composerly focus on the fixed pair of voices can give us reason to favor the dualistic reading.

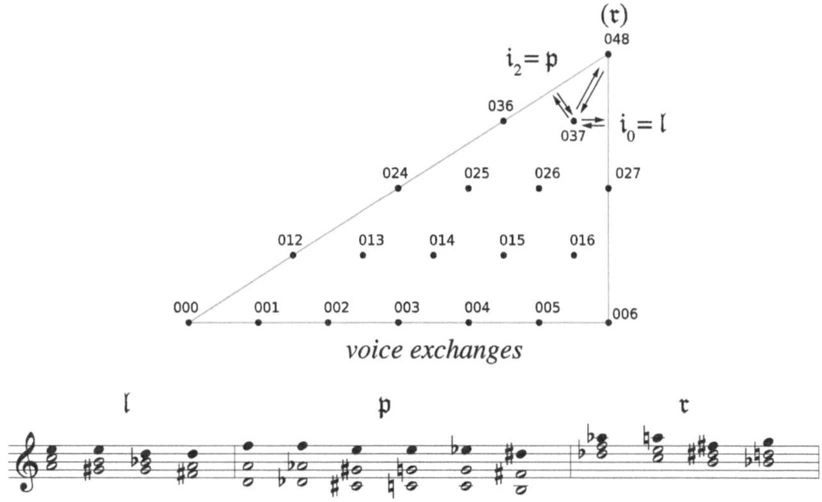

Figure 45. Passages in which two voices descend in parallel while the third alternates between prime and inverted forms: Mozart, C major piano sonata K. 309, I, mm. 73–76, featuring the $\mathfrak{l}$ voice leading; the Benedictus from Schnittke's *Requiem*, featuring $\mathfrak{p}$; and the opening of Gesualdo's *Moro Lasso,* featuring $\mathfrak{r}$.

their starting point. For triads, the $\mathfrak{i}_1$ voice leading bounces off the completely-even augmented triad *E*, but this is not true for other trichords: when a trichord's second-largest interval is not a major third, then $\mathfrak{i}_1$ is represented by paths such as $\mathfrak{i}_2\mathfrak{i}_0\mathfrak{i}_2$ or $\mathfrak{i}_0\mathfrak{i}_2\mathfrak{i}_0$ ($\mathfrak{p}\mathfrak{l}\mathfrak{p}$ or $\mathfrak{l}\mathfrak{p}\mathfrak{l}$), which reflect three times off the two inversional faces.

These sorts of voice leadings can also be found in nontriadic contexts. Figure 46 shows three passages from Stravinsky's *Rite of Spring* that connect 016 triads with neo-Riemannian set-class voice leadings; here we see Stravinsky using two of the three possibilities, preserving the semitone in the first and third examples and the perfect fourth in the second. (Note that here $\mathfrak{i}_1$, or neo-Riemannian $\mathfrak{r}$, is represented by $\mathfrak{i}_0\mathfrak{i}_2\mathfrak{i}_0$.) Figure 47 shows a passage that juxtaposes sonorities that are *approximately* related by inversion, as in Figure 35 above. (The examples in this paragraph were developed in collaboration with Jon Russell, whose remarkable Ph.D. dissertation [Russell 2018] reveals a wealth of intricate voice-leading structure in *The Rite of Spring;* the musical portion of Figure 47 is entirely his.) Figure 48 shows a four-voice version of the technique, moving back and forth along a single path in set-class space and preserving the distance between *two* pairs of voices since the chord's size is even. (Here we shift from a three-voice geometrical picture to a four-voice topological representation.) Unlike the earlier examples, Stravinsky does not simply alternate between inversionally related forms; instead, he moves the top voices in parallel thirds along the C diatonic collection, allowing the quality of the resulting third to determine the set form. Since the lower voices are chromatic transpositions of the upper, the passage's four melodies move along three different diatonic scales.[59] What results is an unusual combination of set-theoretical and contrapuntal thinking, with the former reflected by the music's harmonic consistency (its use of inversionally related set classes) and the latter its traditional focus on efficient voice leading. The simplicity of the graphical analysis aptly reflects the devastating clarity of Stravinsky's musical thinking.

It should be emphasized that inversional set-class space encodes the dualistic view that *ascending* transposition along a normal-form chord is equivalent to *descending* transposition along its inversion, a perspective that is not always analytically appropriate. To the extent that musicians

[59] Furthermore, the passage's underlying voice leading (0, 1, 4, 4) → (0, 0, 3, 4) takes advantage of the inversional symmetry of a two-note set class, as described in Tymoczko (2011, §2.9) (and particularly Figure 2.9.5).

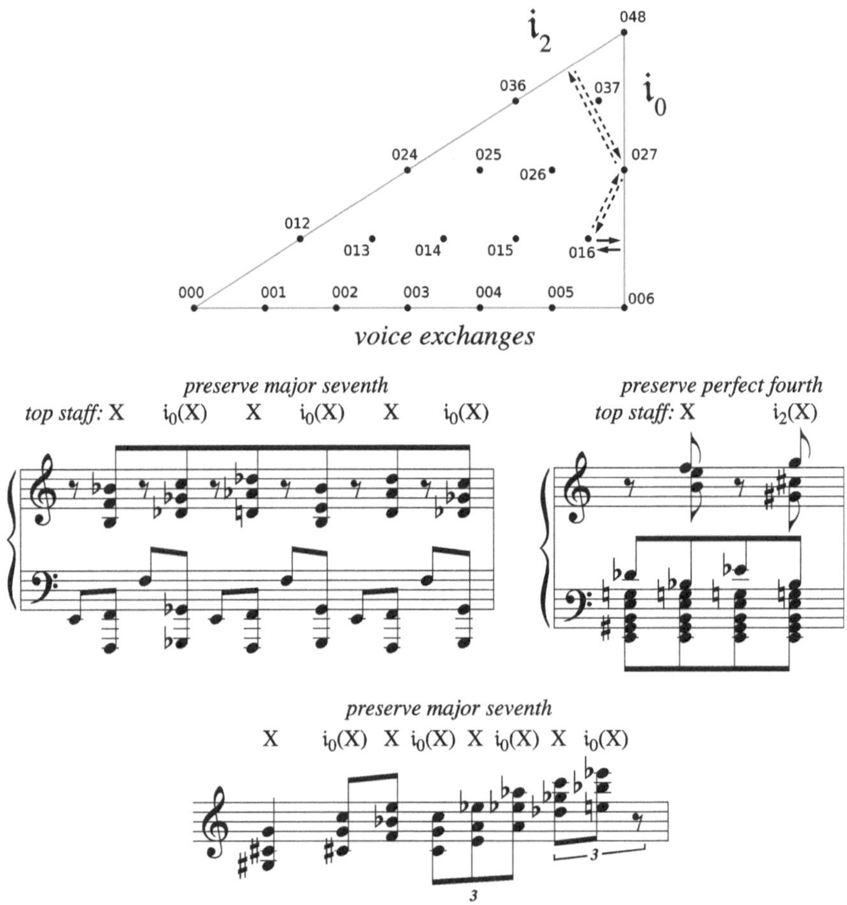

Figure 46. Three passages in *The Rite of Spring* that connect 016 trichords using generalized neo-Riemannian voice leadings: (top left) R114, m. 2 (solid line); (top right) R15, m. 1 (dotted line); (bottom) R64, mm. 5–6 (brass part only, solid line).

think nondualistically (for instance, grouping together ascending transpositions-along-the-chord regardless of a set class's internal structure), the most natural framework is transpositional set-class space, perhaps augmented by the postulate that inversionally related chords are to be considered similar. Fundamentally, the issue is that the (mostly atonal) composers who tend to categorize chords inversionally are least likely to be concerned with efficient voice leading, while the (mostly tonal) composers who are concerned with efficient voice leading are least likely to categorize chords inversionally. This is a significant limitation. Thus while I am convinced that inversional set-class space can be a useful tool for thinking about progressions between inversionally or nearly-inversionally related chords (as in the preceding examples), I am much less confident about its ability to relate *transpositions*.

* * *

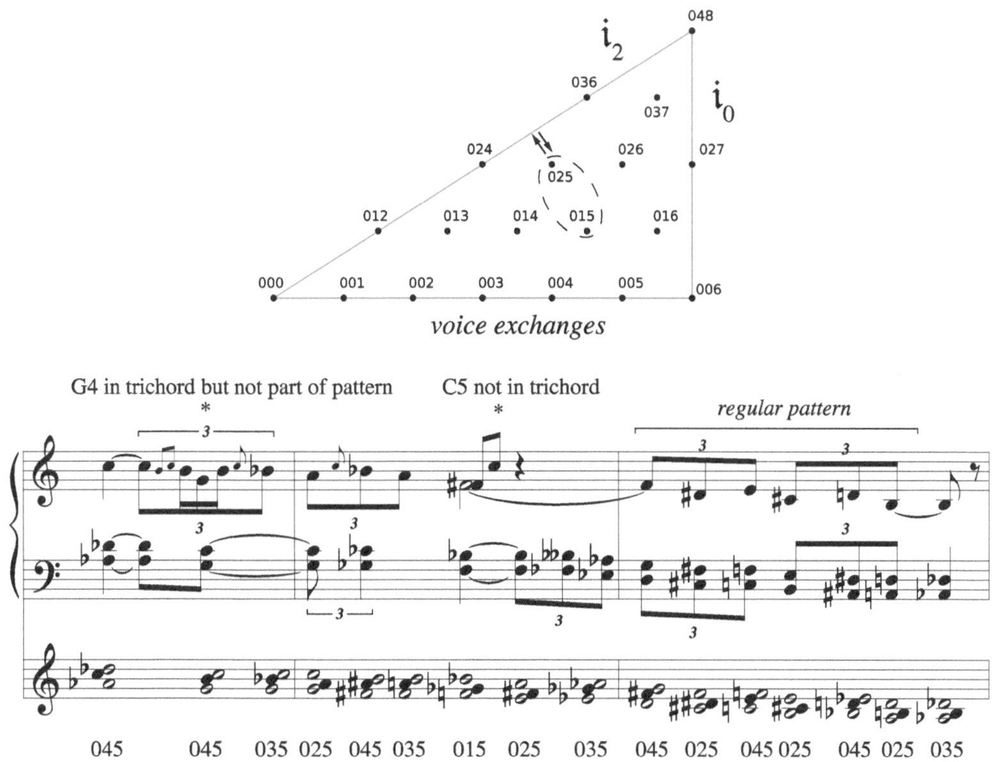

Figure 47. Measures 4–6 present fifteen successive i_2 voice leadings relating 015 and 025 set classes.

Now for the mathematical details. The inversional spaces are defined by an additional constraint on normal-form ordering:

NF4. The second interval in the ordering is less than or equal to the wraparound interval: for a set class $(0, x_1, x_2, \ldots, x_{n-1})$, we require $x_2 - x_1 \leq \boldsymbol{o} - x_{n-1}$, with $\boldsymbol{o}$ the size of the octave.

In transpositional set-class space, the major and minor triads are different set classes with normal forms (0, 3, 8) and (0, 3, 7) respectively; in inversional set-class space they belong to the same equivalence class, with NF4 selecting (0, 3, 7) as the sole normal form.

Geometrically, this replaces one of the vertices of transpositional set-class space, $(0, 0, \boldsymbol{o}, \ldots, \boldsymbol{o})$, with a new vertex $(0, 0, \boldsymbol{o}/2, \boldsymbol{o}/2, \ldots, \boldsymbol{o}/2)$, creating a new simplex half as large as in the transpositional case. The facet representing t_{-1} is unaffected, since its chords have their smallest intervals in their first two positions and hence satisfy NF4.[60] Meanwhile the t_1 facet disappears entirely by virtue of containing points whose minimal interval is in the first and wraparound positions; its replacement contains orderings whose second and wraparound intervals are equal.[61] The remaining facets are subspaces of facets of transpositional space; this is because the original transpositional facets contained both $(0, 0, \ldots, 0)$ and $(0, 0, \boldsymbol{o}, \boldsymbol{o}, \ldots, \boldsymbol{o})$, and the new vertex $(0, 0, \boldsymbol{o}/2, \boldsymbol{o}/2, \ldots, \boldsymbol{o}/2)$ bisects the line connecting them. These relations are spelled out in Figure 49.

[60] This is the facet that neither includes the new vertex nor the vertex it replaced.

[61] The t_1 facet does not contain $(0, 0, \ldots, 0)$; the remaining vertices of set-class space have their second and wraparound intervals equal.

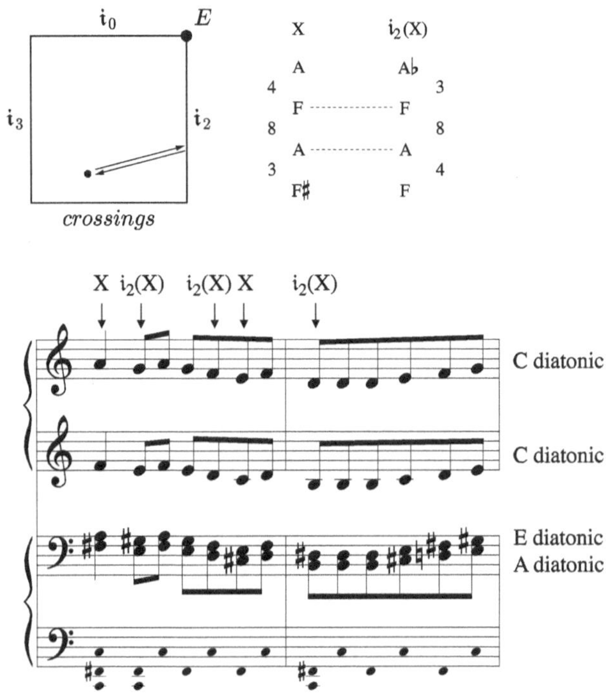

Figure 48. A passage from the *Rite of Spring* involving inversionally related four-note chords; the interval between alto and tenor is always eight semitones; between soprano and bass is fifteen semitones. The entire passage is a repeating series of i_2 voice leadings.

T-space		**I-space**	
vertex	**opposite facet**	**vertex**	**opposite facet**
E	permutation boundary	E	permutation boundary
(0, 0, ..., 0)	t_1	(0, 0, ..., 0)	i_0
(0, 0, $\mathbf{o}$, ..., $\mathbf{o}$)	t_{-1}	(0, 0, $\mathbf{o}$/2, ..., $\mathbf{o}$/2)	i_{-1}
(0, ..., 0, $\mathbf{o}$, ..., $\mathbf{o}$)	t_x	(0, ..., 0, $\mathbf{o}$, ..., $\mathbf{o}$)	i_x

Figure 49. A comparison of the vertices and facets for transpositional and inversional set-class space. Here "opposite facet" means the facet that does not contain the vertex in question.

The interior of the simplex is again banal, recording the divisions of a fixed quantity $\boldsymbol{o}$ into n parts, now subject to the requirements that the first is no larger than any other, and that the second is no larger than the last.[62] This is again a cone whose layers are parameterized by the chord's smallest interval (Figures 50–52). Its base contains set classes with pitch-class duplications, representing voice exchanges as in transpositional space. The remaining boundaries contain pairs of inversionally related set classes as well as inversionally symmetrical chords, with each non-permutational facet containing pairs of set classes related under one particular neo-Riemannian set-class inversion that can serve as its label. For example, if a pair of normal-form n-tuples are related by i_{-1} then their first and second intervals are minimal, which means they lie on the facet that is a subset of the t_{-1} fact of transpositional set-class space; the same is true for i_{-2} and the t_{-2} facet of transpositional set-class space, etc. (Figure 53). Pairs of set classes on these facets are related by the relevant neo-Riemannian set-class inversion; line segments from the interior

[62] Now the office workers are going to n events in different locations, and the smallest car is going to see movie i_0 while the car going to see i_1 is smaller than the car going to see i_{-1}.

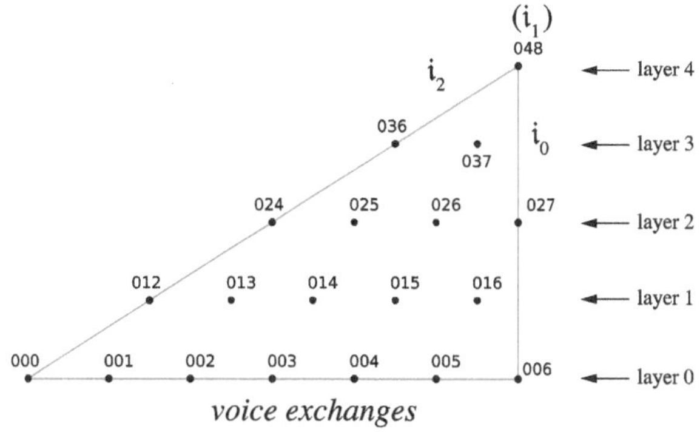

Figure 50. Trichordal inversional set-class space as a cone.

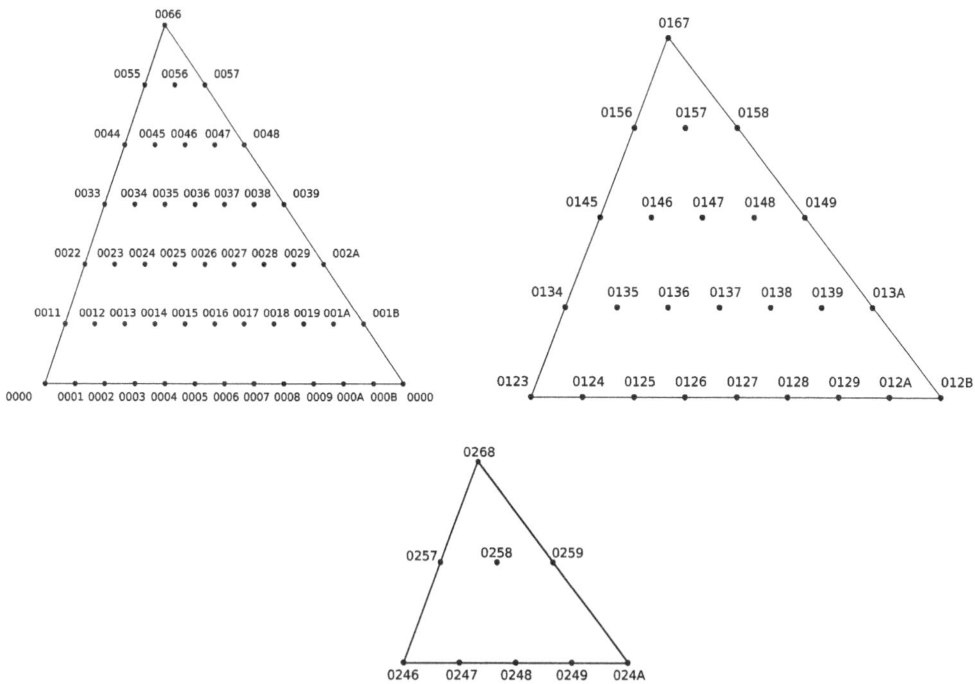

Figure 51. Three layers of tetrachordal inversional set-class space, with smallest interval 0, 1, and 2.

of the simplex can reflect off these facets, returning to their starting point and articulating the relevant inversion. Points on the interior of the simplex cannot be inversionally symmetrical.[63] This explains why we can represent inversional set-class space with the polygons of Figure 44.

There is however an important complication not arising in the transpositional case: higher-dimensional set-class spaces contain features representing not just the inversions but *every other element of the crossing-free voice leading group as well* – which is to say, all of the dualistic transpositions-along-the-chord and the "missing" inversion i_1. This can already be seen in tetrachordal space, which contains linear subspaces acting as t_1, t_2, and t_3 (Figure 54). The

[63] This follows from the fact that they have a unique smallest interval and thus could be symmetrical only under i_0 (by the definition of i_0). But any set class symmetrical under that inversion has its second interval equal to its wraparound interval and hence lies on the i_0 facet of the simplex.

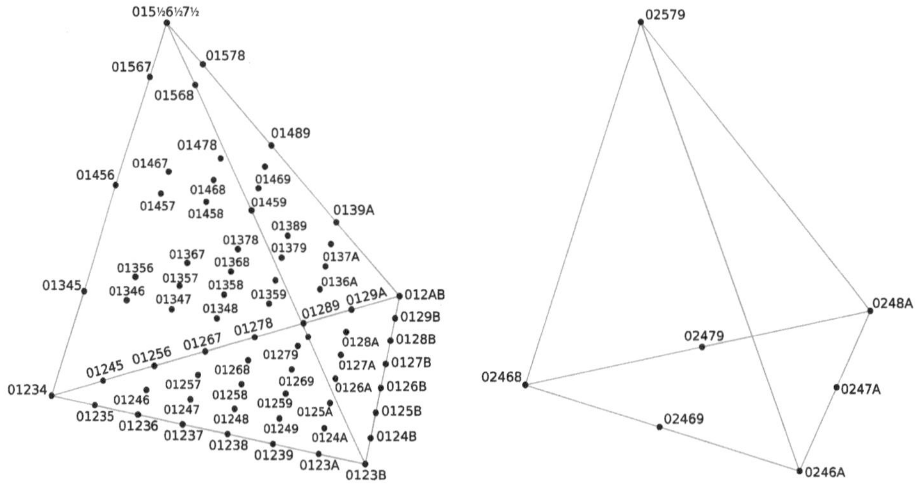

Figure 52. Two layers of pentachordal inversional set-class space, with smallest interval 1 and 2.

i_{-1}								i_{-2}							
x_0		x_1		x_2		x_3		x_4		x_5	
a	*b*	*c*	*d*	*e*	*(f)*			*a*	*b*	*c*	*d*	*e*	*(f)*		
b	*a*	*f*	*e*	*d*	*(c)*			*c*	*b*	*a*	*f*	*e*	*(d)*		
x_2	x_1	x_0	x_5	x_4	x_3			x_3	x_2	x_1	x_0	x_5	x_4		

i_{-1}

$x_0 \quad x_1 \quad x_2 \quad x_3 \quad x_4 \quad x_5$

a b c d e (f)

b a f e d (c)

$x_2 \quad x_1 \quad x_0 \quad x_5 \quad x_4 \quad x_3$

i_{-2}

$x_0 \quad x_1 \quad x_2 \quad x_3 \quad x_4 \quad x_5$

a b c d e (f)

c b a f e (d)

$x_3 \quad x_2 \quad x_1 \quad x_0 \quad x_5 \quad x_4$

Figure 53. If two normal-form orderings are related by i_{-1}, then their first and second intervals (*a* and *b*) must be the same size, and no larger than any other interval. For i_{-2} this is true of the first and third intervals (*a* and *c*).

t_1

$x_0 \quad x_1 \quad x_2 \quad x_3$

a b c d

b c d a

$x_1 \quad x_2 \quad x_3 \quad x_0$

t_2

$x_0 \quad x_1 \quad x_2 \quad x_3$

a b c d

c d a b

$x_2 \quad x_3 \quad x_0 \quad x_1$

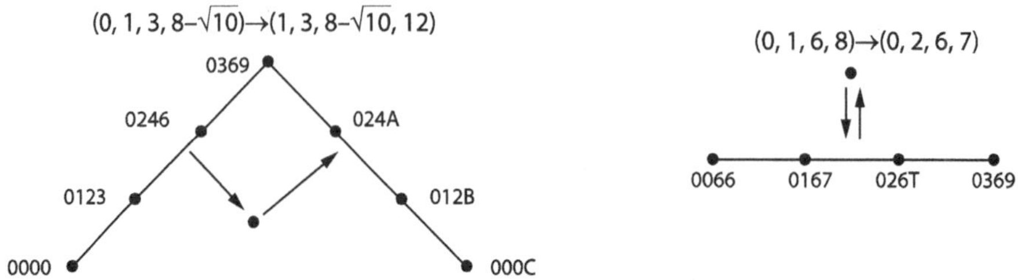

Figure 54. Four-note inversional set-class space contains subspaces representing t_1 and t_2; the former consists of a pair of glued-together line segments, and the latter a mirror-like line segment of t_2-symmetrical chords. There can no subspace representing i_1-invariant set classes, for such chords would have step-interval sequence *abcd* by hypothesis equal to *dcba*, which by NF1–4 implies $a = b = c = d$. The voice leading in the first example, when represented by linear interpolation, passes through the identified line segment; this is the origin of the square root.

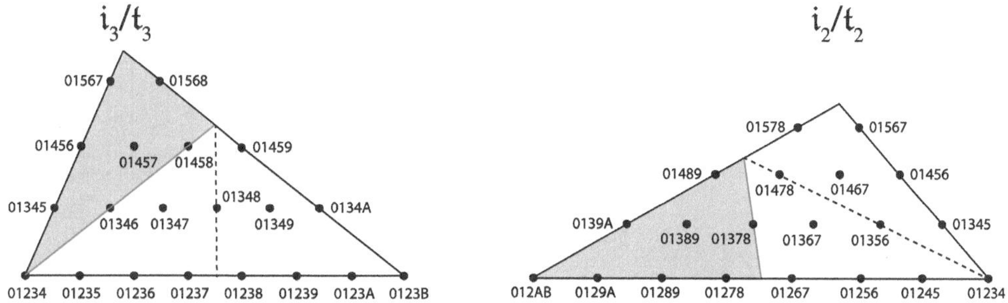

Figure 55. The semitonal layer of the i_2 and i_3 facets of pentachordal space. The shaded triangles are glued together as in transpositional space, representing $t_{\pm 2}$. The lighter areas contain pairs of inversionally related chords, with the dotted line inversionally symmetrical. Every chord on the i_3 facet either has an inversional partner on its facet or a transpositional partner on the i_2 facet: if its normal-form step-interval sequence is *abcde*, then its t_2 and i_3 orderings are *cdeab* and *cbaed* respectively; if $d \leq b$, the t_2 ordering is normal and on the i_2 facet; if $b \leq d$, the i_3 ordering is normal and on the i_3 facet. If $b = d$, then both situations obtain. See the appendix for a more general argument.

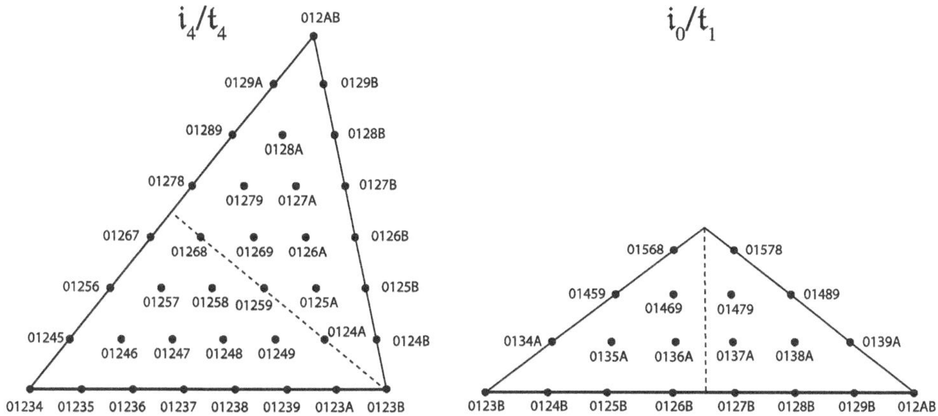

Figure 56. The semitonal layer of the i_4/t_4 and i_0/t_1 facets of pentachordal inversional set-class space. Each facet is divided into in inversionally related halves. However, their bottom edges are glued together and can represent transposition: thus a t_{-1} voice leading can move from the interior of the space to the bottom edge of the i_4 boundary, reappear on the bottom edge of i_0 facet, and return to its starting point, just like the transpositional case.

phenomenon is more dramatic in pentachordal space, with the i_2/t_2 and i_3/t_3 facets containing substantial transpositional regions (Figure 55). Here still, however, the i_4/t_4 and i_0/t_1 facets are almost exclusively taken up with inversionally related pairs, with only a single line segment representing transposition (Figure 56).[64] In higher dimension such facets acquire larger transpositional regions; these spaces also add i_1, which in lower dimension is represented only by *E* (Figure 57). Figure 58 thus provides a more accurate picture of inversional set-class space, with each edge representing *both* a neo-Riemannian set-class inversion and a transposition-along-the chord. – subsuming transpositions-along-the-chord while also adding inversions. (Note that the paired transposition and inversion share the same subscript except in the case of t_1/i_0.) Linear interpolation can produce a variety of different boundary interactions all equivalent to the same basic transformation (Figure 59).

---

[64] The t_{-1} voice leading (0, 0.5, 2, 3, 9.5) $\rightarrow$ (0, 2.5, 3, 4.5, 5.5) is an example of the type of voice leading mentioned in the caption to Figure 56. It moves from the interior of the space to the i_4 boundary at (0, 7/6, 14/6, 21/6, 49/6), corresponding to (0, 1, 2, 3, 7.97) on the semitonal layer, reappears on the i_0 facet at (0, 7/6, 14/6, 42/6, 65/6), or (0, 1, 2, 6.97, 11) on the semitonal layer, and returns to its starting point.

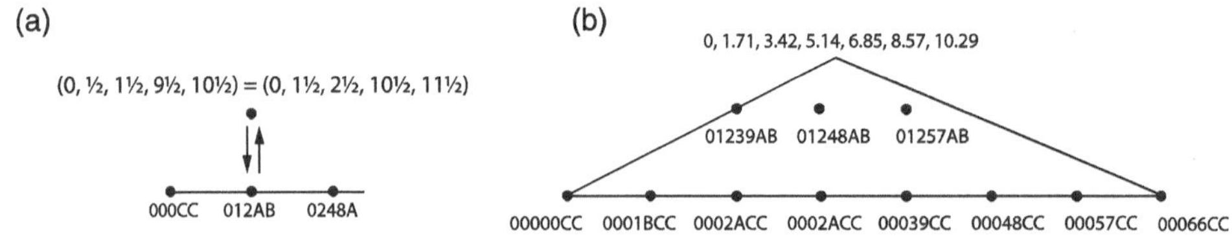

Figure 57. Subspaces invariant under i_1. (a) In pentachordal space, this is a line containing interval cycles. The voice leading on the figure moves from the interior of the space, reflects off the line and returns to itself, acting as i_1. (b) In heptachordal inversional set-class space, there is a two-dimensional subspace of chords symmetrical under i_1, containing sets of the form $(0, a, 2a, 2a + b, 12 - 2a - b, 12 - 2a, 12 - a)$, with $a \leq b$, and $a \leq 12 - 4a - 2b$.

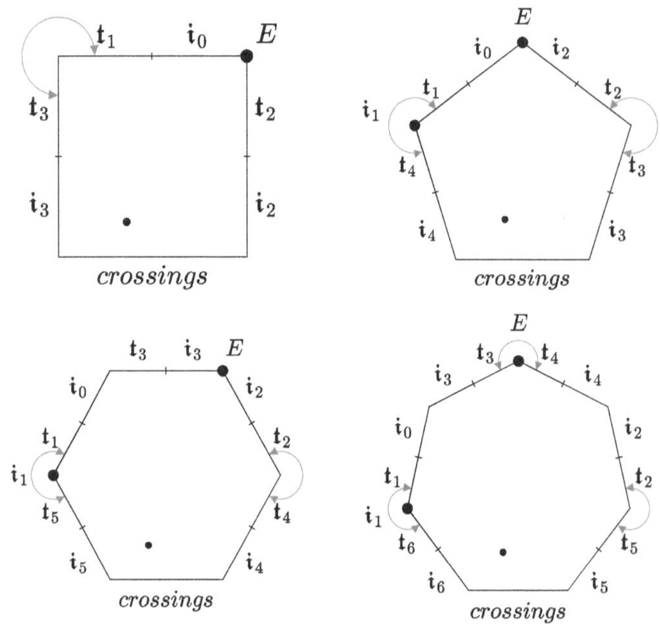

Figure 58. A more accurate picture of inversional set-class space for $n = 4, 5, 6, 7$.

5. Transformation, symmetry, and analytical utility

Geometrical music theory initially used linear interpolation to associate voice leadings with "generalized line segments," relying on visual intuition to convert these paths into analytical observations. This paper instead models voice leadings with homotopy classes of paths, leading to greatly simplified representations. We can go farther by replacing the geometry with an "alphabet" of voice-leading transformations: n pairwise voice-exchanges, n transpositions along the chord, and n neo-Riemannian voice leadings. A path in set-class space spells out a word in this alphabet, the algebraic correlate of a difficult-to-visualize geometrical object. In contexts where we care about specific chords rather than set classes, we can add letters controlling the absolute transpositional level of the destination chord.[65] It is also possible to devise letters that double or eliminate notes in the initial chord, or link unrelated sonorities by a "default" voice leading, or represent the simultaneous crossing of multiple voices.[66]

[65] For instance, let W be any ordered sequence of voice exchanges, transpositions along the chord, and neo-Riemannian voice leadings (c, t, and i as defined below); we can define WS_x as the form of the voice leading whose paths sum to x. For a given starting chord, this is unique.

[66] Here again we touch on ideas similar to those in Roeder (1984).

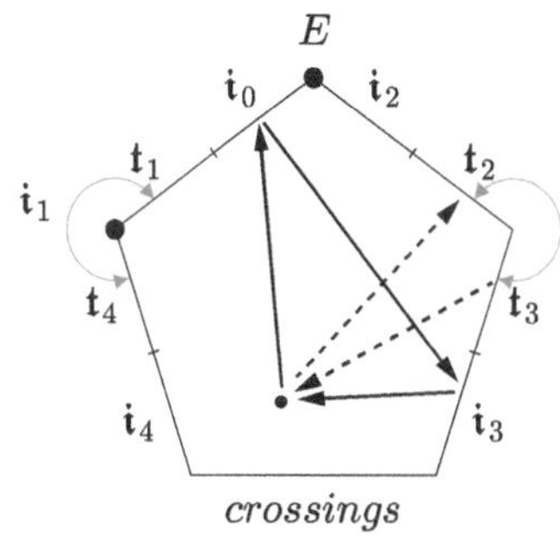

Figure 59. The dotted line is the path traced out by the two-step ascending transposition-along-the-chord (0, 1, 3, 5, 7) → (3, 5, 7, 12, 13); the solid line represents the two-step ascending transposition-along-the-chord (0, 1, 3, 5, 9) → (3, 5, 9, 12, 13). Here linear interpolation produces different patterns of boundary interactions.

Specific transformations can be defined by their action on a set class's normal form, dualistically if we impose NF4 and nondualistically if not.[67] For instance, the top line of Figure 60 defines a voice exchange that swaps the second and third notes in the normal ordering: if we do not use inversion to categorize chords, then the dominant seventh's normal form is (0, 2, 6, 9) and the relevant voice exchange is (0, 2, 6, 9) → (0, 6, 2, 9); if we do, then the dominant seventh's normal form is (0, 2, 5, 8) and we have to invert the voice leading (0, 2, 5, 8) → (0, 5, 2, 8) to obtain the dominant-seventh equivalent (2, 0, 9, 6) → (2, –3, 12, 6) or (0, 2, 6, 9) → (–3, 2, 6, 12) upon reordering.[68] (As the example shows, we can represent the dualistic voice leadings using the same visual pattern of arrows if we write the dominant seventh in descending pitch-class order.) In inversional space the dualistic transformations are more or less forced upon us; but in the other spaces, or when we are using purely algebraic language, we are free to choose.[69]

Figure 61 presents four voice leadings from the prelude to Wagner's *Tristan,* each moving a half-diminished seventh to its inversion, the dominant seventh. The four voice leadings can be labeled $\mathfrak{t}_2\mathfrak{c}_0\mathfrak{i}_2$, $\mathfrak{t}_3\mathfrak{c}_0\mathfrak{i}_2$, $\mathfrak{t}_0\mathfrak{c}_0\mathfrak{i}_2$, and the retrograde of $\mathfrak{t}_1\mathfrak{c}_0\mathfrak{i}_2$.[70] I have written them this way to clarify their structural similarity, indicating that they are all equivalent up to an initial transposition that rotates the destination chord's mode; structurally, each moves the root of the half-diminished seventh to a different note of the dominant seventh, crossing the two voices that sound the dominant seventh's third and fifth. (The fourth voice leading, meanwhile, crosses the Tristan chord's third and fifth.[71]) Our labels provide a structure-neutral description that can be applied to any four-note chord. This is possible because the alphabet mirrors the topological structure of the various musico-geometrical spaces, its elements generating their fundamental groups.[72]

In analytical contexts, the permutation region typically provides the clearest view (§3, Figures 28 and 29), with its n boundaries corresponding to the alphabet's n pairwise voice exchanges,

---

[67] See the appendix. For chords with multiple normal-form orderings we must arbitrarily choose one.

[68] The pairwise voice exchanges exchange two voices pitch classes without crossing any other voices. In the two-note case we need to distinguish c_0, which crosses the closest notes *by the smallest possible paths*, as in (C4, E4) → (E4, C4), from c_1, which crosses the same voices by a longer path, as in (C4, E4) → (E3, C5). For the tritone the distinction between these two voice leadings evaporates as they are the same size.

[69] Readers can use the sc.html website to explore the alphabet.

[70] The retrogrades of $\mathfrak{c}_x$ and $\mathfrak{i}_y$ are $\mathfrak{c}_x$ and $\mathfrak{i}_y$ respectively; the retrograde of $\mathfrak{t}_x$ is $\mathfrak{t}_{-x}$ since reversing an ascending x-step transposition-along-the-chord produces a descending transposition by x steps. Thus the retrograde of $\mathfrak{t}_1\mathfrak{c}_0\mathfrak{i}_2$ is $\mathfrak{i}_2\mathfrak{c}_0\mathfrak{t}_3$ rather than $\mathfrak{i}_2\mathfrak{c}_0\mathfrak{t}_1$.

[71] The inversion of (F, A♭, B, E♭) → (F, B, G, D) around F is (F, D, B, G) → (F, B, E♭, A♭) whose reordered retrograde is (F, A♭, B, E♭) → (F, G, D, B), which is analogous to the other voice leadings in crossing the dominant seventh's third and fifth.

[72] The fundamental group of the permutation region is the group of voice exchanges; for transpositional or inversional set-class space it is the semidirect product of the voice exchanges with either the cyclic group $\mathbf{C}_n$ or the dihedral group $\mathbf{D}_n$. In chord space the transpositions along the chord are instead represented by $\mathbb{Z}$.

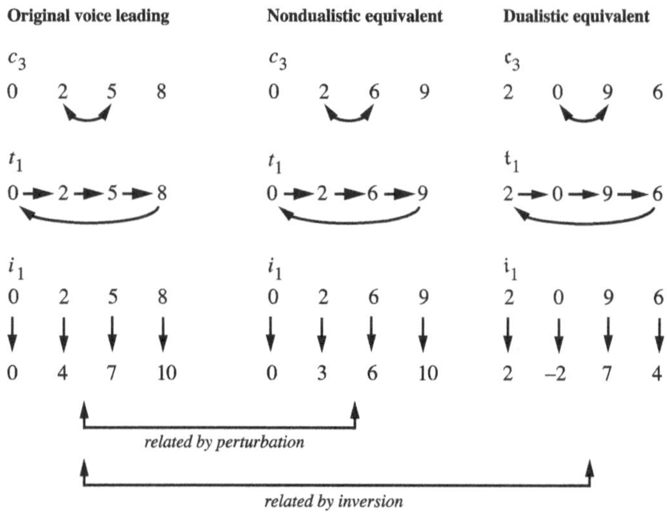

Figure 60. Dualistic and non-dualistic versions of the basic voice-leading transformations. On the left, the transformation expressed with the inversionally normal half-diminished seventh; in the middle, a non-dualistic equivalent; on the right, the dualistic form. The transformations can be expressed by their effect on a normally ordered collection: c_3 exchanges the second and third notes of the normal ordering, t_1 moves each note up by one step along the normal ordering, and i_1 is the crossing-free voice leading between the chord and its inversion that sends the smallest interval down by one order position when the first chord is normally ordered.

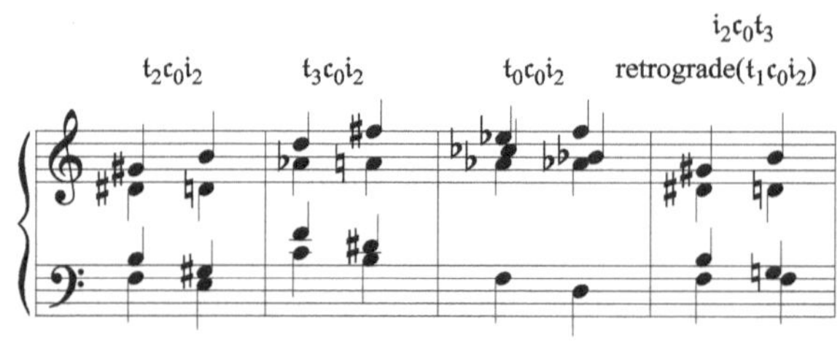

Figure 61. Four voice leadings from the *Tristan* prelude along with their descriptions in the alphabet.

its *n* transpositions the "transpositions along the chord" and the remaining *n* points the neo-Riemannian set-class inversions (Figure 62). Figure 63 uses the permutation region to plot the four *Tristan* voice leadings, with the dark dots representing the four modes of the half-diminished seventh and the lighter dots the dominant seventh.[73] The four voice leadings start at the same point but arrive at different modes of the dominant seventh, collectively exhausting the four possible destinations.[74] Furthermore, the first three reflect off the boundary *opposite* the final dominant seventh, which is the visual representation of the crossing of the dominant's third and fifth. The last voice leading proceeds from its starting point to the opposite boundary and then to its destination chord; it is visually clear that this would be like the others were the direction reversed and light dots exchanged with dark dots – which is to say that it is the *retrograde inversion* of the "missing" fourth path.[75] (Since the dualistic transformations include the inversional relationship by definition, this geometrical observation is equivalent to the algebraic point in

[73] Again, readers can use sc.html website to calculate these paths.

[74] Meyer (1989, 283) argues that the *Tristan* prelude marks the beginning of the decline of the notion of inversional equivalence, but this point seems inconsistent with Wagner's prominent use of inverted Tristan chords, for example in measure 10.

[75] It is the "missing" voice leading insofar as it maps the root of the half-diminished seventh to the seventh of the dominant seventh, whereas the others map root to root, root to third, and root to fifth (ignoring the crossing).

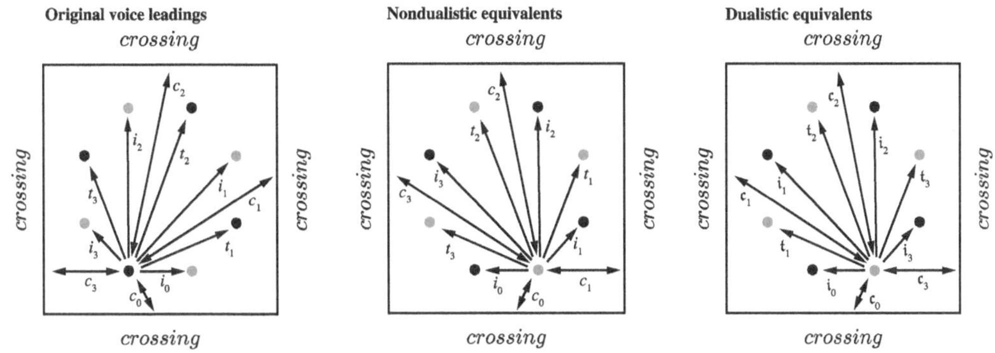

Figure 62. The twelve elements of the basic tetrachordal alphabet in the permutation region. On the left, the inversionally normal half-diminished seventh. In the middle, the non-dualistic versions of these transformations as applied to the dominant seventh. On the right, the dualistic equivalents, reflecting the left figure along its vertical center.

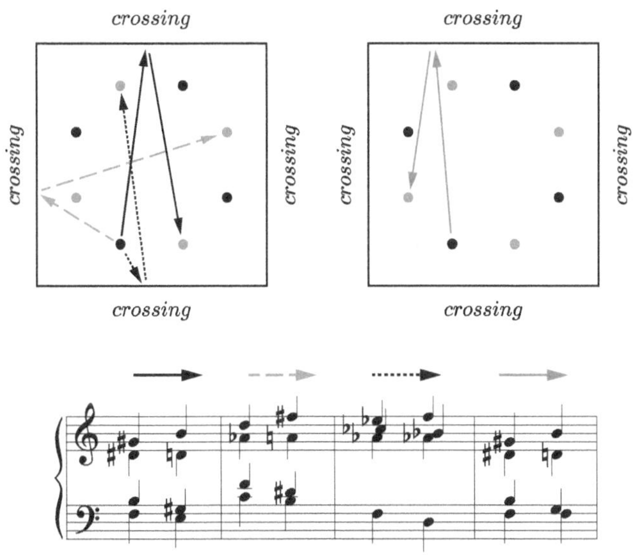

Figure 63. The four *Tristan* voice leadings plotted in the tetrachordal permutation region. The first three touch a crossing boundary and arrive at the opposite dominant-seventh. The last is the retrograde inversion of such a path.

the previous paragraph.) Visual representation thus clarifies the voice leadings' structural logic, allowing us to ask questions like "what would the fourth voice leading look like if it more closely paralleled the other three?" (Answer: it would be the retrograde inversion of the one Wagner actually uses.) Indeed, it was the visual resemblance that first alerted me to the presence of the algebraic relationship.[76]

Such connections are considerably more obscure in the inversional set-class space of Figure 64. The first two voice leadings end similarly, reflecting off the permutation boundary, i_0, and i_2; since i_0i_2 is t_2, the combination can be understood to cross the voices that have the dominant seventh's third and fifth. (Each begins by touching a different inversional boundary that, in combination with i_0i_2, produces a different mode of the dominant seventh.) However, the third voice leading is not obviously similar to the first two, reflecting solely off the crossing and

[76] In *A Geometry of Music* (Tymoczko 2011) I proposed that we could best understand *Tristan* if we chose to disregard voice exchanges; the polygonal models allows us to see that the crossings have an interesting structure as well.

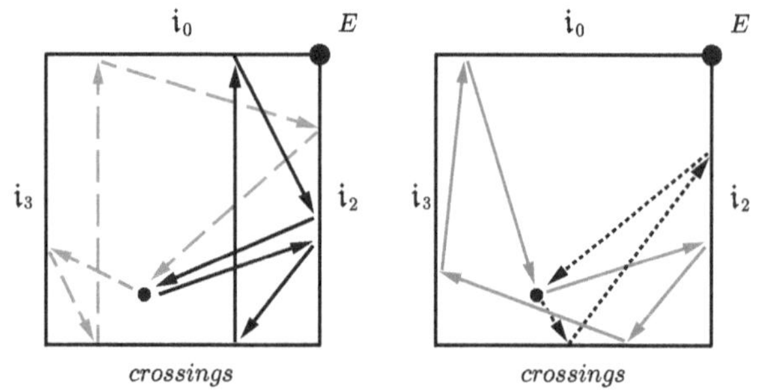

Figure 64. The four *Tristan* voice leadings in inversional set-class space. The paths are much less comprehensible than in the preceding figure.

$\mathfrak{i}_2$ boundaries; instead of three separate inversions, it contains only one. Nor is the retrograde-inversion relationship to the fourth voice leading nearly so obvious: where the first two have the structure $\mathfrak{i}_x\mathfrak{c}\mathfrak{i}_0\mathfrak{i}_2$ (with $\mathfrak{c}$ representing the crossing), the latter is $\mathfrak{i}_2\mathfrak{c}\mathfrak{i}_x\mathfrak{i}_0$. The issue here is that as we move to more abstract geometries, standard musical operations are represented in more and more complex ways, with an ever-increasing distance between geometrical representation and familiar musical categories.[77] Thus c_x in the permutation region (with n copies of each set class) becomes $t_x c_0 \mathfrak{t}_{-x}$ in transpositional set-class space (with just two copies of each set class), and $\mathfrak{i}_0\mathfrak{i}_{-x}\mathfrak{c}\mathfrak{i}_{-x}\mathfrak{i}_0$ in inversional space (with each set class represented just once). For this reason, the permutation region may be the most analytically useful model, with the more abstract spaces better-suited to abstract theoretical questions.

I have argued that the main benefit of the more abstract set-class spaces is to clarify the connection between *efficient voice leading* and *symmetry* (Tymoczko 2006, 2011). Indeed, the three letters in our alphabet, c_x, t_x, and i_x correspond to the three basic symmetries (or near-symmetries) a chord can have. Geometrically this is a matter of being close to singular points in set-class space.[78] The more abstract set-class spaces directly manifest these possibilities in the structure of their boundaries. *Permutational near-symmetry* (or symmetry under c_x) obtains when the set class's notes are close together, which is to say that it is near the permutational boundary. *Inversional near symmetry* or symmetry under i_x, occurs when a chord is near the inversionally symmetrical points dispersed along the i_x boundaries.

Transpositional near-symmetry (or near-symmetry under t_x) comes in two forms. *Transpositional symmetry due to evenness* exploits the conical structure of set-class spaces; nearly even set classes are near the tip of the cone so that all of the transpositions-along-the-chord are small.[79] The other sort of transpositional symmetry exploits some subgroup of the possible transpositions-along-the-chord, represented geometrically by boundaries that are identified with themselves so as to form mirror singularities; thus it is closely analogous to inversional and permutational symmetry, which similarly exploit mirror singularities. For example, an uneven tetrachord can be symmetrical under t_2 by virtue of being close to Figure 33's singular central line; by contrast,

---

[77] The Cohnian *Tonnetz* represents transpositions-along-the-chord as the product of neo-Riemannian inversions, a similarly counterintuitive factoring. In some ways the *Tonnetz* is a chord-based structure reflecting features of inversional set-class space.

[78] A symmetrical chord is invariant under some word W, a combination of c_x, t_y, and i_z, and therefore under the subgroup of voice leadings generated by W; chords near this symmetrical chord can be connected by efficient voice leading to their W-forms.

[79] Here it can be useful to imagine the cone's layers as an $(n - 1)$-sided polygon whose edges represent the non-trivial transpositions along the chord, containing all those set classes whose smallest interval is of a certain size. As the smallest interval increases, this polygon shrinks to a single point. The permutational boundary does not appear in this representation because it is the cone's bottom layer.

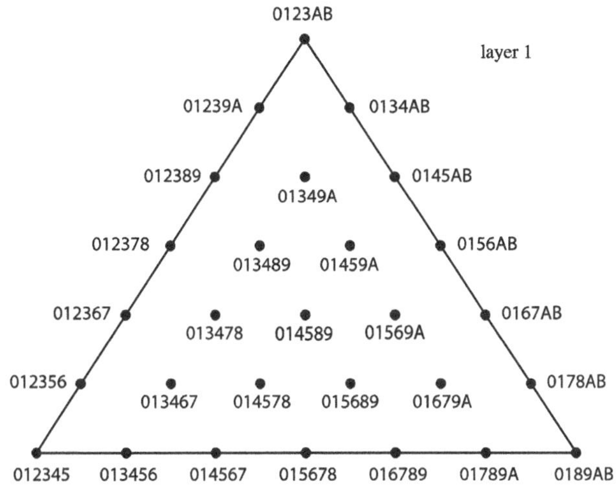

Figure 65. The semitonal layer of the intersection of the t_2 and t_4 boundaries in hexachordal set-class space, containing set classes that remain in normal form under both t_2 and t_{-2} and whose smallest interval appears in the first, third, and fifth positions of the normal ordering. Each set class appears three times and is mapped onto its equivalents by 120° rotation. The central point, meanwhile, is symmetrical under t_2, and hence acts like a mirror in the higher dimensional orbifold. Nearby chords like (0, 0.1, 4, 4.2, 8, 8.3) can be nearly symmetrical under t_2 even though they are close to the base of the cone.

uneven tetrachords cannot be nearly symmetrical under t_1 or t_3 since these boundaries are identified, and hence lack mirror singularities of the appropriate sort. We can find transpositional mirror singularities on all and only those boundaries t_x where x divides the size of the chord and is not equal to 1 or −1 (Figure 65).[80]

Thus we have a hierarchy of increasingly abstract models, ranging from geometrically faithful representations to abstract polygonal diagrams (annular if we focus on chords, polygonal if we focus on set classes) to the completely algebraic letters of our voice-leading alphabet. These are all more or less equivalent, providing different ways to think about the same fundamental relationships. Moreover, within the category of polygonal diagrams there is an important choice between the permutation region, which is more analytically useful, and the more abstract transpositional and inversional set-class spaces, which clarify general theoretical relationships. At the very end of this process, we can represent a vast set of contrapuntal possibilities using an extremely parsimonious alphabet; indeed, the appendix explains that the bijective voice leadings between the transpositions of any n-note chord can be expressed as the repeated application of just three transformations, a single crossing, a single transposition along the scale, and a single transposition along the chord.

6. Conclusion

Though this paper uses mathematical language, musicians in many different genres manage to internalize transformations like "shift a collection by step as if it were a scale," "swap two of a collection's notes," or "invert a chord so as to preserve its spacing in chordal steps." Figure 66, from Mark Levine's *Jazz Piano Book,* shows his catalogue of the five inversions of Bill Evans's "So What" chord. Levine moves each note by step along the notes of the chord, the pentatonic scale C–D–F–G–A, looping counterclockwise through annular space precisely as in Figure 15.

[80] These points can be thought of as recreating features of lower-dimensional set-class space within the more complicated spaces representing larger collections.

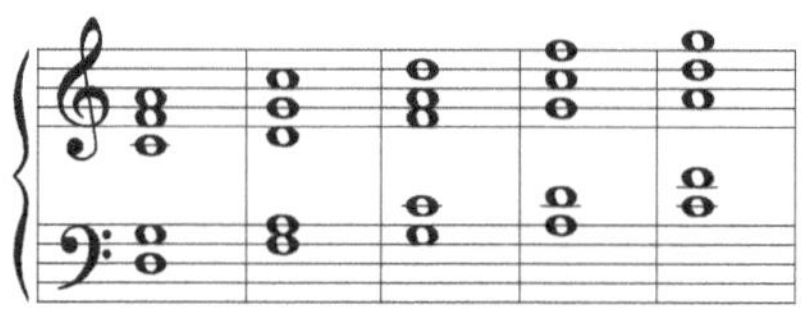

Figure 66. Mark Levine's illustration of the five inversions of the "So What" chord. Each voice ascends by step along the pentatonic scale D–F–G–A–C.

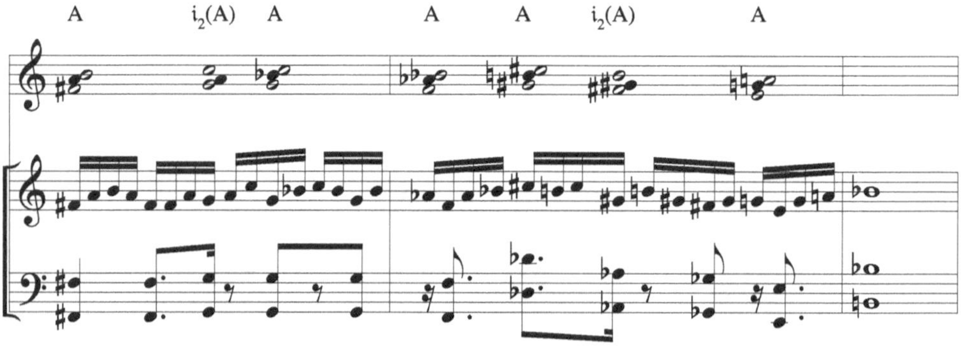

Figure 67. Rudresh Mahanthappa's 2006 composition "The Decider" is based on the 025 set class, moving the perfect fourth in parallel, the middle voice alternately a major second below the top voice and a major second above the bottom. This is very similar to Figures 45 and 46 above.

Figure 67 shows the opening of Rudresh Mahanthappa's "The Decider," articulating a series of inversionally related 025 and 035 sonorities, much like Figures 45–48. For all its theoretical abstraction, the topological approach is eminently practical, its fundamental transformations both compositionally fruitful and musically ubiquitous.

Theorists, however, may find that the shift from geometry to topology requires some conceptual reorientation. When I first started thinking about musical geometry, I wrongly dismissed the neo-Riemannian operations as unnatural (Figure 68). I now realize that a deeper understanding of musical geometry requires us to distinguish two classes of transformation: those representing global symmetries, which allow us to "glue together" regions of musical space, and the *contrapuntal possibilities that remain once we have used these symmetries to create a geometry*, known to musicians as voice leadings and to topologists as the fundamental group. The global symmetries are the *input* to the process of musical modeling, while the contrapuntal possibilities are the *output* of the process, something we learn from the resulting geometry. The difference between them – the difference between traditional transposition and the transpositions-along-the-chord, or between traditional inversion and the neo-Riemannian voice leadings – is a measure of what we can hope to gain from the geometrical approach.

Philosophically, the global symmetries are something you can do *to* a passage of music while the contrapuntal possibilities are, as David Lewin put it, something you can do *inside* a piece of music (Lewin 1987, 159). The shift between these two perspectives was extremely gradual, and in Lewin's work global symmetries are not always distinguished from contrapuntal moves. Schoenberg's 1941 lecture "Composition with Twelve Tones" explicitly connects transposition and inversion to the symmetries of musical space, conceiving them as transformations that can be applied to extended passages such as twelve-tone rows (Schoenberg 1975; these symmetries in turn inspired the five "OPTIC" symmetries in Callender, Quinn, and Tymoczko 2008). The neo-Riemannian transformations, reinterpreted as voice leadings by Cohn, are instead localized moves that can be made inside a particular musical game, and in this sense they are more limited than the global symmetries. But they are no less important, being a crucial first step toward

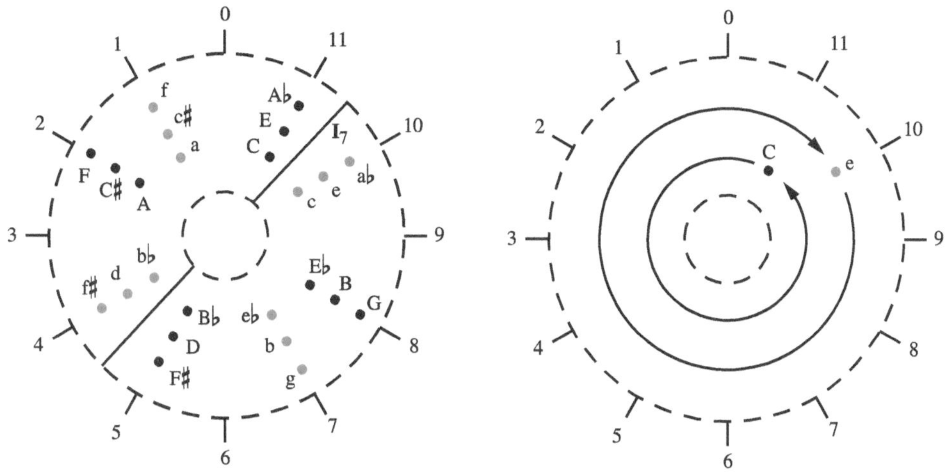

Figure 68. In the annular spaces, traditional transposition and inversion are geometrically straightforward, corresponding to rotation and reflection (here around the solid line). By contrast, dualistic transpositions-along-the-chord move inversions in opposite directions and hence do not correspond to standard geometrical symmetries.

understanding the spaces formed by those global symmetries. This paper has generalized Cohn's voice leadings to arbitrary chords inside arbitrary scales. An even more basic step is the passage from the octave shifts that create pitch-class space to the fundamental group's associated "paths in pitch-class space" – represented geometrically by loops in the pitch-class circle.[81] In higher dimension, these paths become the "transpositions along the chord," creating the annular topology that can be found in countless discrete voice-leading graphs (e.g. Figure 1; Douthett and Steinbach 1998 or §3.11 of Tymoczko 2011). Thus theorists built a piecemeal understanding of the various routes through contrapuntal space, whose totality we can now comprehend as the fundamental group.

It can be challenging to distinguish an orbifold's fundamental group from its symmetry group, as there is a mathematical theorem telling us they are isomorphic (Hughes 2015). This means we are guaranteed to find similar structures serving subtly different musical functions. For musicians the connection between the two kinds of transformation is clearest in the case of voice crossings, which manifestly embody the permutation symmetry giving rise to the space itself while *also* being familiar as voice leadings.[82] By contrast, transpositions-along-the-chord are not at all salient among the global symmetries of chord space even while playing a crucial role in the fundamental group.[83] With inversion the situation is more complex, as the term is used to describe a range of conceptually distinct phenomena, including reflection around a fixed pitch or pitch class (typically a global symmetry), generalized neo-Riemannian voice leadings (paths in voice-leading space, or elements of the fundamental group), and "contextual inversions" having

[81] To form pitch-class space we glue together octave-related points, forming the circular quotient space $\mathbb{R}/\mathbb{Z}$. The elements of the circle's homotopy group, $\mathbb{Z}$, can in turn be interpreted as paths (or voice leadings) moving by some integral number of octaves in pitch class space. When I first started writing about these paths I encountered numerous theorists who claimed they were mathematically ill-defined.

[82] Even here, however, the resemblance can be misleading, as permutation in $\mathbb{R}^n$ is not a voice exchange as I define it, sending (x, y) to (y, x) regardless of content. My c_0 transform instead sends (C4, E5) to (E4, C5), which is not a permutation of pitches; it is also distinct from the c_1 voice exchange, which sends (C4, E5) to (E3, C6).

[83] Two-note chord space is $\mathbb{R}^2/(\mathbf{S}_2 \ltimes \mathbb{Z}^2)$, with the two factors representing permutation ($\mathbf{S}_2$) and octave equivalence ($\mathbb{Z}^2$). The transpositions along the chord do not appear among these global symmetries as their effect depends on the ordering of the dyad: for (C4, E4) it is the first note that moves up by four semitones, while for (E4, C4) it is the second.

something of the character of each.[84] The result can be confusing, with theorists rightly intuiting that there should be "inversion-like" voice leadings even while being unsure how to define them in mathematically or musically fruitful ways.[85] It can be genuinely surprising to find the dihedral group appearing not in its familiar guise as the group of transpositions and inversions (global symmetries), nor in its slightly less familiar guise as the group of *Wechsels* and *Schritts* (contextual inversions and dualistic transpositions), but as a collection of voice leadings linking a set class to itself (the crossing-free subgroup of the fundamental group of set-class space). Here the n-note chord is closely analogous to the $\boldsymbol{o}$-note scale that contains it.

Ironically, by distinguishing global symmetries from the fundamental group, we can start to bridge the divide between the geometrical and transformational approaches to music theory, a divide that is as much philosophical as technical. My own interest in musical geometry reflects a realist methodology, a pursuit of musical constraints that transcend the explicit decisions of any individual composer. By contrast, transformational theory has often been associated with a more relativist philosophy, treating "transformations" as postulates and not attempting to justify them on psychological or other grounds. (These two perspectives can be loosely associated with tonality and atonality, the one language grounded in acoustics, convention, and perhaps biology, the other aspiring to surpass those limits.) This in turn leads to numerous technical divergences, with transformational theory modeling intervals as abstract symmetry groups rather than more concrete geometrical objects such as paths or elements of a tangent space (Tymoczko 2009). Because of this philosophical difference, I was initially slow to realize that a transformational approach to voice leading was closely related to homotopy theory – a lapse which in turn led me to overlook the important fact that boundaries of the simplicial set-class models can be identified with specific voice-leading transformations. The current paper reflects my dawning realization that transformational theory could play a crucial *simplifying* role entirely compatible with a robust realism: we can employ topology without suggesting that all chords are equal, or that any transformation is as good as any other, but simply because we have encountered geometrical structures too complicated to intuit directly.

Acknowledgment

Thanks to Giovanni Albini, Emmanuel Amiot, Julian Hook, Frank Lehman, Jason Yust, and especially Jacob Shulman for helpful suggestions and corrections.

Disclosure statement

No potential conflict of interest was reported by the author(s).

References

Brown, Matthew. 2005. *Explaining Tonality*. Rochester: University of Rochester Press.
Callender, Clifton. 2004. "Continuous Transformations." *Music Theory Online* 10 (3).

[84] There is significant disagreement in the literature about whether neo-Riemannian transformations are contextual inversions, in part because of the way they evolved over time. Earlier work such as Lewin (1987) and Hyer (1989) tends to define neo-Riemannian transformations as contextual inversions, while later work, following Cohn (1996), tends not to. However, some later writers, such as Straus (2011), adopt the former definition while some early writers, like Morris (1998), adopt the latter.

[85] Straus (2003) defines "inversion-like" voice leadings mapping note x in voice A into note $c - x$ in voice A; geometrically these correspond to somewhat unnatural transformations such as $\mathfrak{c}_0\mathfrak{c}_2\mathfrak{c}_0\mathfrak{i}_2$ for trichords and $\mathfrak{c}_0\mathfrak{c}_3\mathfrak{c}_0\mathfrak{c}_2\mathfrak{c}_3\mathfrak{c}_0\mathfrak{i}$ for tetrachords; musically, they are uncommon insofar as they involve a large number of crossings in pitch-class space. Though it is not obvious, Cohn's neo-Riemannian voice leadings are arguably the more natural "inversion-like" voice leadings, being the homotopy group's manifestation of inversional symmetry.

Callender, Clifton, Ian Quinn, and Dmitri Tymoczko. 2008. "Generalized Voice Leading Spaces." *Science* 320: 346–348.
Cohn, Richard. 1996. "Maximally Smooth Cycles, Hexatonic Systems, and the Analysis of Late-Romantic Triadic Progressions." *Music Analysis* 15 (1): 9–40.
Cohn, Richard. 1997. "Neo-Riemannian Operations, Parsimonious Trichords, and their '*Tonnetz*' Representations." *Journal of Music Theory* 41 (1): 1–66.
Cohn, Richard. 2003. "A Tetrahedral Model of Tetrachordal Voice Leading Space." *Music Theory Online* 9 (4).
Cohn, Richard. 2011a. *Audacious Euphony*. New York: Oxford.
Cohn, Richard. 2011b. "Tonal Pitch Space and the (Neo-)Riemannian *Tonnetz*." In *The Oxford Handbook of Neo-Riemannian Music Theories*, edited by Alex Rehding, and Ed Gollin, 322–348. New York: Oxford University Press.
Douthett, Jack, and Peter Steinbach. 1998. "Parsimonious Graphs: A Study in Parsimony, Contextual Transformations, and Modes of Limited Transposition." *Journal of Music Theory* 42 (2): 241–263.
Fiore, Thomas M., and Ramon Satyendra. 2005. "Generalized Contextual Groups." *Music Theory Online* 11 (3).
Genuys, Grégoire. 2019. "Pseudo-distances Between Chords of Different Cardinality on Generalized Voice-Leading Spaces." *Journal of Mathematics and Music* 13 (3): 193–206.
Haimo, Ethan. 1996. "Atonality, Analysis, and the Intentional Fallacy." *Music Theory Spectrum* 18 (2): 167–199.
Hook, Julian. 2008. "Signature Transformations." In *Music Theory and Mathematics: Chords, Collections, and Transformations*, edited by Jack Douthett, Martha M. Hyde, and Charles J. Smith, 137–160. Rochester: University of Rochester Press.
Hook, Julian. 2011. "Spelled Heptachords." In *Mathematics and Computation in Music, Proceedings of the Third International Conference of the Society for Mathematics and Computation in Music*, edited by Carlos Agon, Emmanuel Amiot, Moreno Andreatta, Gérard Assayag, Jean Bresson, and John Mandereau, 84–97. New York: Springer.
Hughes, James. 2015. "Using Fundamental Groups and Groupoids of Chord Spaces to Model Voice Leading." In *Mathematics and Computation in Music, Proceedings of the 5th International Conference*, edited by Tom Collins, David Meredith, and Anja Volk, 267–278. New York: Springer.
Hyer, Brian. 1989. "Tonal Intuitions in *Tristan und Isolde*." Ph. D. dissertation, Yale University.
Lerdahl, Fred. 2001. *Tonal Pitch Space*. New York: Oxford University Press.
Lerdahl, Fred. 2020. *Composition and Cognition*. Berkeley: University of California Press.
Lewin, David. 1987. *Generalized Musical Intervals and Transformations*. New Haven: Yale University Press.
Meyer, Leonard. 1989. *Style and Music*. Chicago: University of Chicago Press.
Morris, Robert. 1998. "Voice-Leading Spaces." *Music Theory Spectrum* 20 (2): 175–208.
Regener, Eric. 1974. "On Allen Forte's Theory of Chords." *Perspectives of New Music* 13 (1): 191–212.
Roeder, John. 1984. "A Theory of Voice Leading for Atonal Music." Ph. D. dissertation, Yale University.
Roeder, John. 1987. "A Geometric Representation of Pitch-Class Series." *Perspectives of New Music* 25 (1–2): 362–409.
Russell, Jonathan. 2018. "Harmony and Voice Leading in the Rite of Spring." Ph. D dissertation, Princeton University.
Schoenberg, Arnold. 1975. *Style and Idea: Selected Writings of Arnold Schoenberg*. Edited by Leonard Stein, Translated by Leo Black. New York: St. Martins Press.
Siciliano, Michael. 2002. "Neo-Riemannian Transformations and the Harmony of Franz Schubert." Ph.D. dissertation, The University of Chicago.
Simon, Herbert. 1956. "Rational Choice and the Structure of the Environment." *Psychological Review* 63 (2): 129–138.
Straus, Joseph. 2003. "Uniformity, Balance, and Smoothness in Atonal Voice Leading." *Music Theory Spectrum* 25 (2): 305–352.
Straus, Joseph. 2007. "Voice Leading in Set-Class Space." *Journal of Music Theory* 49: 45–108.
Straus, Joseph. 2011. "Contextual Inversion Spaces." *Journal of Music Theory* 55 (1): 43–88.
Stuckenschmidt, Hans Heinz. 1965. "Debussy or Berg? The Mystery of a Chord Progression." *The Musical Quarterly* 51 (3): 453–459.
Tymoczko, Dmitri. 2006. "The Geometry of Musical Chords." *Science* 313: 72–74.
Tymoczko, Dmitri. 2008. "Scale Theory, Serial Theory, and Voice Leading." *Music Analysis* 27 (1): 1–49.
Tymoczko, Dmitri. 2009. "Generalizing Musical Intervals." *Journal of Music Theory* 53 (2): 227–254.
Tymoczko, Dmitri. 2010. "Geometrical Methods in Recent Music Theory." *Music Theory Online* 16 (1).

Tymoczko, Dmitri. 2011. *A Geometry of Music*. New York: Oxford University Press.
Tymoczko, Dmitri. 2012. "The Generalized *Tonnetz*." *Journal of Music Theory* 56 (1): 1–52.
Tymoczko, Dmitri. 2016. "In Quest of Musical Vectors." In *Mathemusical Conversations: Mathematics and Computation in Performance and Composition*, edited by Elaine Chew, Gérard Assayag, and Jordan Smith, 256–282. Singapore: Imperial College Press/World Scientific.
Tymoczko, Dmitri, and Aditya Sivakumar. 2018. "Intuitive Musical Homotopy." In *Mathematical Music Theory*, edited by Robert Peck, and Maria Montiel, 233–252. Singapore: World Scientific.

Mathematical Appendix

Unless otherwise noted all definitions can be found in Tymoczko (2011), Callender, Quinn, and Tymoczko (2008), and Tymoczko (2006).[86]

The basic setting is the continuous circular *pitch-class space* whose points are ordered by log-frequency modulo the octave. A *scale* is a collection of $\boldsymbol{o}$ distinct pitch classes defining a piecewise linear metric with the distance between adjacent notes equal to one ("the scale step"); we use this metric to label pitch classes with integers, choosing an arbitrary element as pitch class 0 and proceeding in the ascending direction. In principle, pitch classes need not lie within the scale, though in most musical applications they will. A one-note scale gives the limiting case of an undivided pitch-class space. Every scale defines a notion of addition and hence rotation (musical "transposition") and reflection (musical "inversion"). Typically I choose the octave to have either size 12, with 0 = C, 1 = C♯, etc., or size 7, with 0 = C, 1 = D, etc.

As described in the paper, a *path in pitch class space* is an ordered pair (p, r) whose first element is a pitch class and whose second element is a real number representing how the note moves, as measured in scale steps. Neither p nor r need be integers. Paths in pitch-class space can be associated with points in the tangent space of the pitch-class circle; each path connecting a pitch class to itself can be identified with a unique element of the circle's fundamental group. A path (p_1, r_1) is *connected* to (p_2, r_2) if it ends where the second begins, $p_2 = (p_1 + r_1) \bmod \boldsymbol{o}$. A *voice leading* is a multiset of paths in pitch-class space; it links two chords, or multisets of notes, $X = \{p_i\}$ and $Y = \{(p_i + r_i) \bmod \boldsymbol{o}\}$.[87] The notation $(x_0, x_1, \ldots, x_{n-1}) \to (y_0, y_1, \ldots, y_{n-1})$ refers to the voice leading $\{(x_i, y_i - x_i)\}$. I sometimes use real numbers or pitches to notate pitch-class voice leadings, e.g. writing (0, 4) → (–8, 12) or (C4, E4) → (E3, C5) to indicate the voice leading in which pitch class C (in any octave) moves down by eight semitones while E (in any octave) moves up by eight semitones. When I use octave-free names, as in (C, E, G) → (C, F A), the assumption is that each voice moves to its destination along the shortest possible path. Voice leadings can be transposed and inverted in the obvious way: transposition rotates a voice leading's pitch classes while leaving its real numbers unchanged; inversion inverts the pitch classes while multiplying the real numbers by –1. Two voice leadings represent the same *transpositional voice-leading schema* if they are related by transposition. A *concatenation* of two voice leadings v_1 and v_2 is a bijection between their paths such that each path in v_1 is connected to its partner in v_2.

It is often useful to add the real components of a collection of paths. A *voice exchange* is a bijective voice leading from a chord to itself whose paths' real numbers sum to zero. A *transpositional voice leading* by x scale steps is a voice leading whose real numbers are all equal to x. A voice leading is *strongly crossing-free* if its voices never cross no matter how they are arranged in register. Strongly crossing-free voice leadings preserve a chord's spacing as measured in chordal steps. A voice leading that is not strongly crossing free has *voice crossings in pitch-class space.* Define *path extension* as the operation of adding the same constant to each of the real numbers in each of a voice leading's paths: two voice leadings are related by *individual transposition* if they are related by a combination of transposition and path extension; they are related by *individual inversion* if they are related by a combination of inversion and path extension. *Voice leadings in transpositional set-class space* are equivalence classes of voice leadings related by individual transposition; *voice leadings in inversional set-class space* are equivalence classes of voice leadings related by individual transposition or individual inversion.

Let $(x_0, x_1, \ldots, x_{n-1})$ be a multiset of pitch classes in normal order as defined by either NF1–3 or NF1–4. The one-step ascending transposition along the chord t_1 is

$$(x_0, x_1, \ldots, x_{n-1}) \to (x_1, x_2, \ldots, x_0 + \boldsymbol{o})$$

The one-step descending transposition along the chord is its retrograde, with n-step transpositions generated by iterating these single-step operations. The pairwise voice exchanges c_i cross only adjacent voices:

$$(\ldots, x_i, x_{i+1}, \ldots) \to (\ldots, x_{i+1}, x_i, \ldots), i < n-1$$
$$(x_0, \ldots, x_{n-1}) \to (x_{n-1} - \boldsymbol{o}, \ldots, x_0 + \boldsymbol{o}), i = n-1$$

[86] In Tymoczko (2006) I used traditional pitch-class intervals rather than paths in pitch-class space as I was afraid to abandon music-theoretical orthodoxy.

[87] Every voice leading is a bijective voice leading between the two multisets X and Y as just defined. A *nonbijective* voice leading is one in which the chords' doublings are not fixed in advance, e.g. (C, C, E, G) → (A, C, F, F), which is a bijective voice leading from (C, C, E, G) to (F, F, A, C) but a nonbijective voice leading from (C, E, G) to (F, A, C).

the contrapuntal design, with numbers corresponding to the paths in each voice:

$$\{((\mathrm{A},0), \quad (\mathrm{A},0), \quad (\mathrm{A},12), \quad (\mathrm{A},0), \quad (\mathrm{A},0)),$$
$$((\mathrm{A},0), \quad (\mathrm{A},0), \quad (\mathrm{A},0), \quad (\mathrm{A},-12), \quad (\mathrm{A},0))\}$$

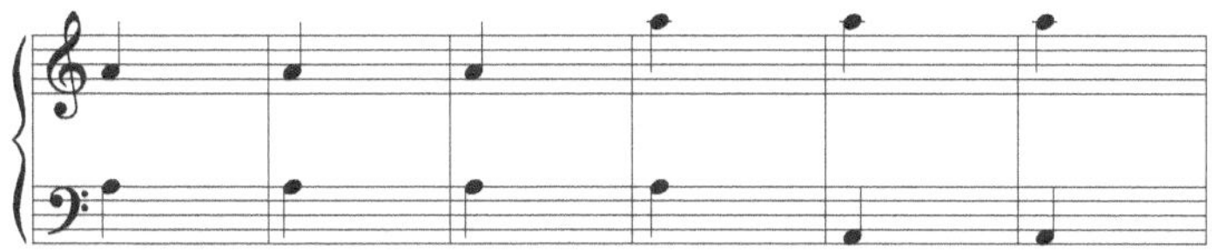

I or c_0 *I or* c_0 $t_1, t_1c_0,$ *or* c_0t_1 $t_{-1}, t_{-1}c_0,$ *or* c_0t_{-1} *I or* c_0

(A4, A3) → (A4, A3) → (A4, A3) → (A5, A3) → (A5, A2) → (A5, A2)

Figure A1. A passage of music containing voice leadings from the singular (A, A) to itself. This passage defines a unique melodic design, but can be described in multiple ways using elements from the orbifold fundamental group.

The neo-Riemannian inversion i_0 is

$$(x_0, x_1, \ldots, x_{n-1}) \to (-x_1, -x_0, \boldsymbol{o} - x_{n-1}, \boldsymbol{o} - x_{n-2}, \ldots, \boldsymbol{o} - x_2)$$

with the remaining neo-Riemannian inversions i_i defined by $t_i i_0$ and sending note x_0 to the inversion of x_{1-i}. (When working with chords rather than sets, it is typical to define more specific transformations that combine these neo-Riemannian inversions with an additional transposition.) If normal forms are defined using NF1–3 only, then we have non-dualistic transformations notated with lowercase: t_i, c_i, and i_i, to which we can add T_i for transpositions along the scale (transpositional voice leadings as defined above). For "dualistic" transformations defined with NF1–4, I use the same labels in fraktur font: $\mathfrak{t}_i$, $\mathfrak{c}_i$, and $\mathfrak{i}_i$, with uppercase $\mathfrak{T}_i$ for transpositions along the scale (a notation not appearing in this paper).

Every n-note chord corresponds to a unique point in the orbifold $\mathbb{T}^n/\mathbf{S}_n$, with the symmetric group $\mathbf{S}_n$ permuting the n circular coordinates. This "symmetric product space" can be described as the twisted product of a simplex and a circle, with its circular dimension coordinatized by the sum of the chord's pitch classes modulo $\boldsymbol{o}$. (Equations defining a fundamental domain are given below.) The space's boundaries are singular and contain chords with note duplications. For a non-singular chord X (i.e. one in the interior of the space with n distinct pitch classes) there is an isomorphism between the group of bijective voice leadings $X \to X$ (combining by the concatenation of connected paths) and the orbifold fundamental group with basepoint X (Hughes 2015). This is true even though voice leadings are discrete objects while the fundamental group represents collections of continuous paths; in general, continuous glissandi provide an intuitive model of the discrete objects, helping us understand what voice leading corresponds to a given path and vice versa.

For a singular chord S (i.e. one with pitch-class duplications), there are multiple ways to concatenate the voice leadings $S \to S$. The orbifold fundamental group at basepoint s, described in Hughes (2015), includes musically unobservable operations that recreate the same fundamental group found at the non-singular basepoints. In musical contexts this is problematic for reasons that are best explained by example. Suppose you were to put your fingers on C4 and F4; there are four basic voice leadings that collectively generate *all* the voice leadings to any other C, F pair: two voice exchanges, c_0 or (C4, F4) → (F4, C4) and c_1 or (C4, F4) → (F3, C5), moving voices in contrary motion by 5 and 7 semitones respectively, and two one-step transpositions along the chord, t_1 or (C4, F4) → (F4, C5) and its retrograde t_{-1} or (C5, F4) → (F4, C4). If on the other hand you place your fingers on A3 and A4, then the small voice exchange c_0 is no longer distinguishable from the identity, while each of the remaining transformations can be applied in two different ways: for c_1 you have both (A3, A4) → (A4, A3) and (A3, A4) → (A2, A5), for t_1 you have (A3, A4) → (A4, A4) and (A3, A4) → (A3, A5), and for t_{-1} you have the retrogrades of the t_1 voice leadings.

From the musician's standpoint this is a significant difference; in the one case there are four possibilities while in the other there are six. The orbifold fundamental group identifies the two situations by postulating an unobservable permutation c_0 (A3, A4) → (A3, A4) that is distinct from the identity (A3, A4) → (A3, A4). The analytical problem with this strategy is illustrated by Figure A1, which presents a succession of voice leadings from the multiset {A, A} to itself. Underneath the music I show that each voice leading is systematically ambiguous depending on whether we choose to include the unobservable permutation c_0 or not. (When voices change octaves there is the additional freedom to put the voice crossing before or after the transposition, which for a nonsingular chord generates distinct voice leadings.) The result is a proliferation of indistinguishable analyses, all arising from the imposition of the same group structure at singular and nonsingular basepoints.

Hughes has suggested that this redundancy is the price we have to pay to model voice exchanges. But my own work uses an alternative strategy that seems to me more musical, namely specifying how voice leadings are to be concatenated. This can be accomplished by modeling an n-voice, m-chord passage as a *contrapuntal design,* or multiset of m-tuples whose elements form a sequence of paths in pitch-class space, each connected to its successor, with each m-tuple representing a distinct melodic line.[88] As with voice leadings themselves, these lines are specified only by their sequence of pitch-class paths, rather than by instrument, register, etc. In the case of non-singular chords, a contrapuntal design is just a sequence of concatenated voice leadings; when singularities are involved, the contrapuntal design adds structure by stipulating how the concatenation is to occur.

The top of Figure A1 shows the unique contrapuntal design corresponding to the passage, determined by the voices as expressed by musical notation. In general, two sequences of orbifold fundamental-group elements map to the same contrapuntal design if and only if they define the same sequence of paths-in-pitch-class-space in each voice, precisely as in the non-singular case. For this reason there is no increased music-analytical power resulting from the additional structure in the orbifold fundamental group, over and above what can be expressed using contrapuntal designs: contrapuntal designs collect musically indistinguishable group-theoretical descriptions into analytically relevant categories. In this sense, the theory of voice leading extracts the musically relevant portion of orbifold homotopy theory.[89] Since these complications are largely tangential to this paper, as well as to the practical concerns of music making, I will ignore them in what follows, assuming non-singular chords for the sake of expositional simplicity.

The orbifold fundamental group for an n-note chord X is $\mathbb{Z} \ltimes (\mathbf{S}_n \ltimes \mathbb{Z}^{n-1})$ with $\mathbb{Z}$ the subgroup of strongly crossing-free voice leadings and $\mathbf{S}_n \ltimes \mathbb{Z}^{n-1}$ the subgroup of voice exchanges (Tymoczko and Sivakumar 2018). The voice exchanges are bijective voice leadings from a chord to itself whose paths have real components summing to 0; these combine a permutation $\mathbf{S}_n$ with a set of octave displacements $\mathbb{Z}^{n-1}$, with the superscript n–1 because one displacement is determined by the sum-0 condition (Hughes 2015). Voice exchanges are generated by the pairwise voice exchanges c_i defined above. The crossing-free subgroup $\mathbb{Z}$ acts by repeated transposition-along-the-chord t_i and contains the transpositional voice leadings $n\mathbb{Z}$ as a normal subgroup. The quotient $\mathbb{Z}/n\mathbb{Z}$ is the cyclic group of transpositions-along-the-chord in set-class space and one of the primary objects discussed in this paper; it is the crossing-free subgroup of the fundamental group of transpositional set-class space. (In all the set-class spaces, the subgroup of voice exchanges is $\mathbf{S}_n \ltimes \mathbb{Z}^{n-1}$, just as in chord space.) This cyclic group is in turn the rotational subgroup of the dihedral group corresponding to the generalized neo-Riemannian voice leadings, the crossing-free subgroup of the fundamental group of inversional set-class space, generated by combining the voice leading $\mathfrak{i}_0$ with the dualistic transpositions along the chord $\mathfrak{t}_i$.

Thus each of the main categories of voice leading is associated with a different group: the scalar transpositions with the cyclic group, the neo-Riemannian set-class inversions with the dihedral group, and the voice exchanges with the symmetric group (augmented by octave shifts summing to zero). The bijective voice leadings between the transpositions of any n-note chord in any $\boldsymbol{o}$-note scale, regardless of cardinality, can in fact be generated by repeated application of just three transformations, a single voice-exchange c_0, the one-step ascending transposition along the chord t_1, and the one step descending transposition along the scale T_{-1}. This is because t_1 and T_{-1} combine to form all the combinations of transposition along the chord and transpositions along the scale, and because $\mathfrak{t}_i c \mathfrak{t}_{-i}$ produces the pairwise voice exchanges.[90] For transpositional set-class space we need only c_0 and t_1, while for inversional space we need $\mathfrak{c}_0$, $\mathfrak{t}_1$, and $\mathfrak{i}_0$.

A fundamental domain for $\mathbb{T}^n/\mathbf{S}_n$, can be obtained by conjoining NF2 in the paper with

$$\mathrm{NF1}'.\ 0 \leq \Sigma x_i \leq \boldsymbol{o}$$

In the orbifold, the sum-0 facet is identified with the sum-$\boldsymbol{o}$ facet; the circular dimension's coordinate is given by Σx_i mod $\boldsymbol{o}$. Since the space's only singularities are on the boundary, its interior is a manifold. The previous paragraph's crossing-free subgroup $\mathbb{Z}$ is isomorphic to the fundamental group of this interior. This isomorphism has a natural musical interpretation: two continuous paths in the interior are homotopic if they represent glissandi whose net effect is to move every pitch class by the same number of semitones, and they can be identified with the unique voice leading whose paths move those same pitch classes by those same amounts. That is, every homotopic glissando moves pitch class p_i by r_i semitones in total, with $\{(p_i, r_i)\}$ the associated voice leading. This isomorphism motivates the annular spaces in §2. The construction can be straightforwardly extended to include voice leadings between distinct (non-singular) chord types and voice leadings passing through the orbifold's singularities (that is, voice leadings with voice crossings in pitch-class space), though these last are difficult to represent visually.

A fundamental domain for the permutation region in §3 can be obtained by combining NF2 with the constraint $\Sigma x_i = k$ for some constant k. For transpositional set-class space, or $\mathbb{T}^{n-1}/\mathbf{S}_n$, we add NF3. A detailed description of these spaces (including their conical structure, which plays a significant role in this paper) is given in Callender, Quinn, and Tymoczko (2008). The main argument of §3 connects all but one of the "boundaries" of transpositional set-class space to the nontrivial transpositions-along-the-chord (here, $\mathbb{Z}/n\mathbb{Z}$). The argument proceeds by describing the facets of the simplicial fundamental domain for set-class space, showing that these facets are identified in the appropriate way, and explaining why paths in the quotient space are associated with glissandi having the appropriate effect. This is another application of the basic postulate of voice-leading geometry, linking voice leadings to equivalence classes of glissandi, only now in a space that ignores transpositions. The reasoning is straightforward in large part because of the choice of

---

[88] While this formal definition is new to this paper, the underlying strategy is implicit in many of my previous analyses, where I often write, e.g. (F, C) → (A, A) → (C, F) to indicate that one voice moves F–A–C while the other moves C–A–F (for an example see Tymoczko 2016, 269).

[89] In set-class space we do sometimes need the additional structure provided by orbifold homotopy theory to account for the highly singular perfectly even chord (§3).

[90] For example t_{-1} is $t_{n-1}\mathrm{T}_{-\boldsymbol{o}}$ while T_1 is $t_n\mathrm{T}_{1-\boldsymbol{o}}$.

fundamental domain; with a different choice the connection between boundaries, homotopy classes, and voice leadings would be much less clear.

The argument in §4 largely parallels that of §3 but with the addition of NF4, augmenting the group $\mathbb{Z}/n\mathbb{Z}$ with the neo-Riemannian set-class voice leading $\mathfrak{i}_0$, acting as described in the text. Once again, the paper associates the boundaries of the simplicial fundamental domain with transformations. The reasoning is more involved than in the transpositional case because a single boundary of the fundamental domain can represent both a transposition along the chord and a neo-Riemannian inversion. This is shown, in the lower-dimensional cases, by direct construction. To see that the phenomenon occurs in higher dimensions, consider a normal-form point X with step intervals $(s_0, s_1, \ldots, s_i, s_{i+1}, \ldots, s_{n-1})$, with $s_1 \leq s_{n-1}$ by NF4. Let $s_0 = s_i$ so that X is on the $1, i+1$ facet of inversional set-class space. Now consider the i-step transposition-along-the-chord $\mathfrak{t}_i(X) = (s_i, s_{i+1}, \ldots, s_{n-1}, s_0, s_1, \ldots, s_{i-1})$. If $s_{i+1} \leq s_{i-1}$ then this ordering is normal and we have a pair of transpositionally related boundary points X, $\mathfrak{t}_i(X)$. Alternatively, if $s_{i+1} \geq s_{i-1}$ then $\mathfrak{i}_i(X) = (s_i, s_{i-1}, \ldots, s_1, s_0, \ldots, s_{i+1})$ is normal and on the $1, i+1$ facet, giving us a pair of inversionally related boundary points X, $\mathfrak{i}_i(X)$.

None of these ideas are complicated by the standards of contemporary mathematics, nor are any of the relevant proofs challenging: mostly it is a matter of choosing musically and mathematically appropriate definitions and understanding familiar geometry, particularly that of quotient spaces and the standard simplex. The interest, in my view, lies in the many musical consequences that flow from this simple mathematical picture, leading to new technological tools both literal and metaphorical. To my mind this reflects music theory's status as an applied discipline: where mathematicians are more likely to be interested in generality and proof, music theorists need a detailed and intuitive understanding of the specific spaces arising in musical contexts.

Generalized *Tonnetze* and *Zeitnetze*, and the Topology of Music Concepts

Jason Yust

The music-theoretic idea of a *Tonnetz* can be generalized at different levels: as a network of chords relating by maximal intersection, a simplicial complex in which vertices represent notes and simplices represent chords, and as a triangulation of a manifold or other geometrical space. The geometrical construct is of particular interest, in that allows us to represent inherently topological aspects to important musical concepts. Two kinds of music-theoretical geometry have been proposed that can house *Tonnetze*: geometrical duals of voice-leading spaces and Fourier phase spaces. Fourier phase spaces are particularly appropriate for *Tonnetze* in that their objects are pitch-class distributions (real-valued weightings of the 12 pitch classes) and proximity in these space relates to shared pitch-class content. They admit of a particularly general method of constructing a geometrical *Tonnetz* that allows for interval and chord duplications in a toroidal geometry. This article examines how these duplications can relate to important musical concepts such as key or pitch height, and details a method of removing such redundancies and the resulting changes to the homology of the space. The method also transfers to the rhythmic domain, defining *Zeitnetze* for cyclic rhythms. A number of possible *Tonnetze* are illustrated: on triads, seventh chords, ninth chords, scalar tetrachords, scales, etc., as well as *Zeitnetze* on common cyclic rhythms or timelines. Their different topologies – whether orientable, bounded, manifold, etc. – reveal some of the topological character of musical concepts.

1. Definitions of *Tonnetze*

For the purpose of this article, I will define a *Tonnetz* very generally according to the maximum-overlap or common-tone principle, following the suggestion made in a well-known article by Cohn (1997). However, it will be useful to think of *Tonnetze* in turn at different levels of abstraction, as networks of pitch-class sets, as topological objects, or as geometrical objects. The network construct is the most general, the geometrical one the most specific. In addition, while the definitions are set out initially with pitch-class sets in mind (subsets of the 12-note scale understood as integers mod 12), they may also be applied to beat-class sets, looped rhythms understood as subsets of some periodically repeating set of evenly spaced time points. We will therefore construct *Tonnetze* in any possible *universe*, *u*, where this can refer to any equal division of an octave or of a time cycle. Beat-class sets bring other possible universes into play in a way that may lead to a wider range of applications than only considering other possible equal divisions of the octave. When generalized this way, the term "*Tonnetz*" is not entirely appropriate, so I will call this a "*Zeitnetz*."[1] All of the following definitions may be modified to apply to *Zeitnetze* by replacing "pitch class" everywhere by "beat class." The two constructions are mathematically indistinguishable.

We begin from what is usually referred to as the "dual *Tonnetz*," in which nodes represent chords rather than tones, which I will call the *Tonnetz graph.*

Definition 1.1 A cardinality-n *Tonnetz graph* is a connected graph whose vertices correspond to pitch-class sets of cardinality n in some universe u, such that two vertices may share an edge only if their corresponding sets share $n-1$ elements in common, and the entire network maps onto itself under any transposition of universe u.

The last criterion, transposability, implicitly requires that, for any pitch-class set in the network, all of its transpositions must also be in the network.[2] However, it allows for the network to contain multiple distinct set classes. The other defining feature is that the connections reflect only the maximum-overlap relationships. The definition does allow for a possible maximum-overlap relationship to be excluded, because there may be geometric reasons for such an exclusion, as we will see. However, the connectivity condition may often require that all available connections must be included, in which case we can deduce a unique *Tonnetz* graph given just the set types (transposition-types) it contains. This also means that the choice of set types is not entirely arbitrary, because there must be a way to relate every type to every other through a series of maximum-overlap relationships.

Tonnetz graphs are common currency in much recent mathematical music theory, and therefore their theoretical value appears to be well recognized. What is often described as the dual of the standard triadic *Tonnetz*, or "chicken-wire torus" (Douthett and Steinbach), is a *Tonnetz* graph. The standard *Tonnetz* only involves one chord type and its inversion (major and minor triads) and this is often assumed to be an essential property for generalizing the *Tonnetz* – in that sense the above definition is unusually broad. Assuming only one set class, however, is too restrictive when we move beyond the case of three-note sets. The definition of *Tonnetz* graphs also encompasses a wider range of networks that have been proposed such as those of Douthett and Steinbach (1998) and Žabka (2014), all of which are *Tonnetz* graphs. Similarly, Tymoczko's (2011, 2012) "chord-based networks" are *Tonnetz* graphs. In much of this work (particularly in Tymoczko's), an additional constraint is put upon the networks that adjacent chords must be close in a voice-leading sense. That is, the unique non-common elements must be a "step" apart in pitch-class space (one semitone, one or two semitones, or a generic scale-step, depending on the context). That restriction is lifted here, but we will find that under certain conditions it emerges naturally.

Historically a *Tonnetz* is a network of tones, where chords are represented by groups of connected pitch classes. By mapping the chords in a *Tonnetz* graph to simplices in a network of tones, we can define a *simplicial Tonnetz.*

Definition 1.2 A cardinality-n *simplicial Tonnetz* is a simplicial complex whose nodes are pitch classes in some universe u, such that a *Tonnetz* graph is given by mapping n-simplices to nodes and connecting nodes that overlap in $(n-1)$-simplices, and such that every node and edge belong to some n-simplex participating in this mapping.

---

[1] London (2002) suggests using this term for a different kind of network, but his definition has not entered into common usage.

[2] An invertibility condition similar to the transposability condition might also be added to this definition, so that the *Tonnetz* graph must include all representatives of each T_nI-type set class. All the *Tonnetze* and *Zeitnetze* appearing as examples below would satisfy such a condition.

Generalizations of the *Tonnetz* as a simplicial complex have been studied by Bigo et al. (2013, 2015), Cantazaro (2011), and Giavitto and Spicher (2016) under somewhat more general definitions than the one proposed here, which fixes the cardinality of the maximal simplices and inherits transposability from the *Tonnetz* graph. It is clear that a *Tonnetz* graph can map to a simplicial *Tonnetz* in a canonical way, and vice versa, so the simplicial *Tonnetz* is essentially the same level of generalization as the *Tonnetz* graph. However, geometrical realizations of generalized simplicial *Tonnetze* are a more specific construction that has been less studied (examples include Tymoczko 2012 and Yust 2018a).

Definition 1.3 A cardinality-*n* *geometrical Tonnetz* is an embedding of a cardinality-*n* simplicial *Tonnetz* in some *n*-dimensional geometric space, such that the pitch classes are each associated with unique points in the space, all transpositions correspond to rigid transformations of the space, and the regions defined by the *n*-simplexes are disjoint and cover the space – i.e. they constitute a triangulation of the space.

The geometrical *Tonnetz* realizes the transposability condition for *Tonnetz* graphs geometrically, as rigid transformations. Since the number of simplices is finite, the space must be compact. The shift from the simplicial to geometrical *Tonnetz* is momentous where musical interpretation is concerned, because through a musical interpretation of the geometry we may associate specifically geometrical constructs such as paths and regions with musical concepts. Two significant ways of doing this have been proposed: through voice leading (Callender, Quinn, and Tymoczko 2008; Tymoczko 2012) and through harmonic quality (Quinn 2006; Amiot 2013, 2016; Yust 2015b) as defined by the discrete Fourier transform on pitch-class sets. With the latter type of geometry, a very general method of construction is defined in Yust (2018a) that can embed any simplicial Tonnetz in a geometric space, specifically a *phase space*. To illustrate the main points of interest – the two ways of defining musical geometries, the methods of folding, and the music-conceptual significance of each of these – we begin by working carefully through the specialized cases with a high degree of redundancy, generated collections. These methods are then applied to two-interval generated collections, which include a number of examples of great musical interest (including the standard triadic *Tonnetz*).

2. Generated collections

2.1. *Dyadic Tonnetze*

At the extreme of intervallic redundancy are generated collections, sets generated by multiple iterations of a single interval. Generated collections are special in that they are accurately captured by a *Tonnetz* graph in the form of a plain cycle graph with a single set type. This is true of all dyadic *Tonnetze*, which are trivially generated. So, for example, a *Tonnetz* graph for ic1 dyads would look like:

. . . — {B,C} — {C,C♯} — {C♯,D} — {D,E♭} — {E♭,E} — . . .

Its associated simplicial *Tonnetz* is a one-dimensional cyclic space, whose edges correspond to the nodes of the *Tonnetz* graph, overlapping in individual vertices:

. . . . — B — C — C♯ — D — E♭ — . . .

Similarly a *Tonnetz* graph of perfect fourths/fifths,

. . . — {C,G} — {G,D} — {D,A} — {A,E} — . . .

has an associated simplicial *Tonnetz*:

$$\ldots — \mathrm{C} — \mathrm{G} — \mathrm{D} — \mathrm{A} — \mathrm{E} — \ldots$$

Recall that at this level, the *Tonnetze* are understood as simplicial complexes – in this case made up of 1-simplices or edges. To turn them into geometrical *Tonnetze* is essentially an act of interpretation, assigning a geometrical interpretation to the circle, in this case the pitch-class circle and the circle of fifths.

One interpretation is offered by the application of the discrete Fourier transform (DFT) to pitch-class sets, which is discussed at length elsewhere (Quinn 2006; Amiot 2013, 2016; Yust 2015a, 2015b, 2016, 2017a): these can be understood as *phase spaces*, where the position on the cycle represents the phase value of a certain Fourier component, denoted $\mathrm{Ph}_{k/u}$ for component k in universe u, or simply Ph_k where the universe is understood (e.g. for pitch-class sets, $u = 12$). It is not necessary to re-introduce all of the mathematics of the Fourier transform here: for present purposes, we can understand each Ph_k as an interval cycle in the sense of pitch-class set theory (Perle 1977; Straus 2016), where Ph_1 is the integers mod u and Ph_k is given by multiplying the pitch-class numbers by k mod u. For consistency with previous publications (reflecting the standard definition of the DFT) Ph_1 is treated throughout as the inverse of the pitch-class circle, so that ascent by 1 in Ph_1 corresponds to descent by semitone (or going back by one beat class). Other pitch-class sets, or multisets or weighted pitch-class sets, can then be located in the phase space by taking circular averages of their constituent pitch classes. This gives the space a robust geometry: points in-between pitch classes have specific meaning. In the circle of fifths, or Ph_5, for example, the point halfway between C and G represents the C-G dyad.[3] To get a dense topology, we consider not only regular pitch-class sets, but also weighted ones, AKA pitch-class *distributions*. Weightings may have multiple musical applications, a common one being to represent the probability or frequency of appearance of a pitch class within a certain timespan (Huron 2006; Temperley 2007). The phase spaces provide a one-to-one complete list of ways of arranging the pitch classes that are true to the group-theoretic properties of the transposition group $\mathbf{Z}_{12}$ (or more generally, $\mathbf{Z}_u$) in the sense that the pitch classes map uniquely to points in the space, and transposing them corresponds to some rigid transformation of the space.

Another way to give geometrical meaning to these cycles is through Callender, Quinn, and Tymoczko's (2008) idea of voice-leading spaces. We might first note that the pitch-class circle is the basic one-dimensional voice-leading space, the space of log-frequency modulo the octave.[4] This is a different interpretation than Ph_1: a point half-way between C and C♯, e.g., represents a C quarter-sharp rather than the dyad {C,C♯}. However, this interpretation is not really native to the theoretical universe of the *Tonnetz*, where the simplices are usually taken to represent something like chords or collections, because the ic1 dyads in the one-dimensional voice-leading space represent segments of the log-frequency continuum, not dyadic collections. A more appropriate derivation from voice-leading geometries begins from the two-dimensional voice-leading space. By taking the geometric dual of this space, as proposed by Tymoczko (2012), we get points that correspond to individual pitch classes. Here it is more appropriate to consider the second example above, the circle of fifths. Each point in the space represents a line in the dyadic voice-leading space, the line corresponding to all voice leadings that hold a particular pitch class constant in one voice while moving the other voice. Motion between these represents a rotation of this line around a fixed dyad (some perfect-fifth dyad). This is illustrated on the left side of Figure 1. The

[3] This relationship, however, is not one-to-one. The point representing {C,G} in Ph_5 also represents {F,D}. The relationship only becomes one-to-one if we consider *all* possible phase spaces $0 \leq k \leq u/2$ and the magnitudes of each of these Fourier coefficients.

[4] Throughout, "voice-leading spaces" will refer to what Callender, Quinn, and Tymoczko (2008) specifically call "OP-space." That is, it is voice-leading space folded to recognize octave and permutational equivalences.

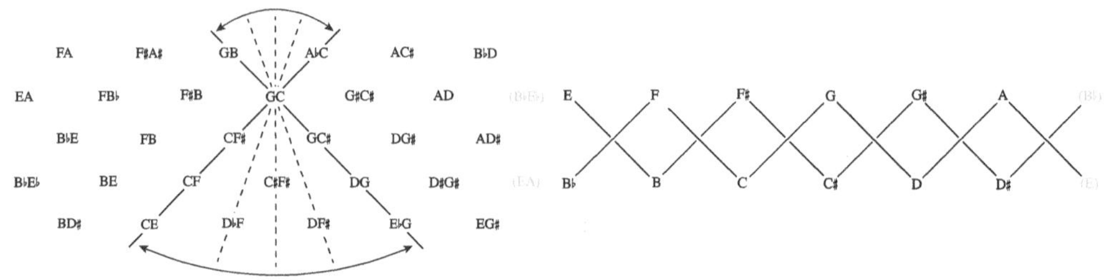

Figure 1. Rotation around a perfect-fifth dyad in voice-leading space, and the *Tonnetz* space as the geometrical dual of dyadic voice-leading space.

point half-way between C and G on the circle of fifths then corresponds to a different line, one that gives a wedge voice leading (or *balanced* voice leading) around the perfect-fifth dyad.

The circle of fifths is not actually smooth according to this interpretation. The dual of the dyadic voice-leading space is shown on the right of Figure 1. Points in this space are lines in the voice-leading space (with pitch classes again indicating lines that hold a pitch class constant), and motion to the left and right here corresponds to sliding the line along the axis of transposition. Perpendicular to this (up and down) are rotations of the line around the point where they cross the center of the space (the tritone axis). Diagonal lines combine both of these kinds of motion, equivalent to rotating around some other fixed point. These spaces are dual in the sense that lines in the *Tonnetz* space correspond to points in the voice-leading space just as lines in the voice-leading space correspond to points in the *Tonnetz* space. The circle of fifths is a zig-zag line that circles the *Tonnetz* space twice, changing direction each time it hits a pitch class. A similar zig-zag line could be defined for the ic1 intervals, but it would be an extremely inefficient way to move through the space, circling almost the entire space for each dyad, and crossing multiple other ic1 lines.[5] Following the line of transposition, on the other hand, would give a very efficient representation of the pitch-class circle, but does not actually reflect the idea of holding a pitch class constant as we go from one ic1 dyad to the next, which is essential to the idea of the *Tonnetz* as a network of common-tone relationships. Voice leading therefore prioritizes certain possible *Tonnetze* over others in a way that phase spaces do not.

Tymoczko (2012) does not actually consider the *Tonnetz* of Figure 1 as a candidate for a dyadic voice-leading space, because its two-semitone voice leadings are not minimal. Note that when the line segment corresponding to a fifth in the dual space (the right side of Figure 1) rotates from one fifth to another, it passes through the tritone. For example, between {F,C} and {C,G} is {F♯,C}. We can also see this in the voice-leading space itself. According to the logic of voice leading, then Figure 1 presents an incomplete, one might even say inaccurate, picture. Voice leading from one fifth to another with a common tone *necessarily* passes through an intermediate tritone, and the voice leading {F,C}→{B,F♯} is the same size, or smaller, than {F,C}→{C,G}, not six times as large, as it is in our *Tonnetz*. For this reason, Tymoczko only considers the possibility of *Tonnetze* that include the most even chord types (such as tritones) before including the next-most even and so forth. This highlights a crucial difference with Fourier phase spaces. In Fourier spaces, the tritone is not necessarily an intermediary between the two fifths; this would be true in Ph_2-space, which reflects, in a certain sense, the logic of dyadic voice leading. But it is not true in Ph_5. The intermediary between {F,C} and {C,G} in *all* phase spaces, including Ph_5, is the shared pitch class C.

[5] A common-tone voice leading between ic1 dyads that is equally efficient to the one on ic5 dyads is in fact possible (moving one note by two semitones at each stage) but it involves a voice crossing. Stated with respect to the dual space, we can say that edges can only rotate through a vertical axis (corresponding to the wedge voice leading) not the horizontal axis (corresponding to transposition). Rotations of the latter type correspond to a voice crossing.

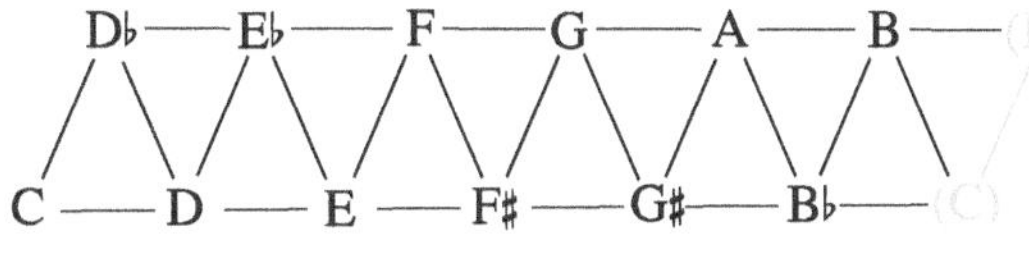

Figure 2. Simplicial *Tonnetz* of chromatic trichords.

Whether one includes tritones or not, the voice-leading interpretation gives geometrical meaning to a *Tonnetz*, but does not technically satisfy the definition of a geometrical *Tonnetz* given in Section 1, because the regions corresponding to the perfect-fifth 2-simplexes are not disjoint: tritone-related perfect fifths cross at the wedge voice leading that relates them. If the size of the universe was odd (hence sharing no factors with 2), this would not occur and voice leading would give a good *Tonnetz* geometry on maximally even dyads (for instance, fifths in a generic-diatonic, $u = 7$, universe).

Because of the relative simplicity of phase spaces as a geometrical realization of *Tonnetze* (with pitch classes and higher-cardinality chords in the same space rather than dual spaces) and because the relationship of phase spaces to interval cycles gives this interpretation a valuable level of generality, we will use phase spaces for the remainder of this section and return to voice leading in Sections 4 and 5.

2.2. *Generated trichordal Tonnetze*

As we increase the cardinality of the generated collection, we can continue to use the same form of *Tonnetz* graph. For instance, the chromatic trichord has a graph:

$$\ldots — \{C,C\sharp,D\} — \{C\sharp,D,E\flat\} — \{D,E\flat,E\} — \{E\flat,E,F\} — \ldots$$

and a chromatic tetrachord:

$$\ldots — \{C,C\sharp,D,E\flat\} — \{C\sharp,D,E\flat,E\} — \{D,E\flat,E,F\} — \{E\flat,E,F,F\sharp\} — \ldots$$

and so forth. A simplicial complex of chromatic trichords might look like Figure 2. Despite a simple cyclic *Tonnetz* graph, the simplicial *Tonnetz* requires a second dimension, bounded by the ic2s, the intervals unique to a given chromatic trichord, because a 2-simplex (triangle) requires a space of a minimum of two dimensions. To embed this simplicial complex in a geometry raises the question of what this second dimension might represent musically.

One way to address this question is a procedure I suggest in Yust (2018a): begin by defining a *Tonnetz* for the chromatic trichord in a two-dimensional phase space that has duplications of ic1, and then eliminate the duplications by defining a folding, resulting in a space like Figure 2 that inherits geometric meaning in the bounded dimension from the original phase space. A trichordal Tonnetz can be embedded in a two-dimensional phase space Ph_{k_1,k_2} (the direct product $\mathrm{Ph}_{k_1} \times \mathrm{Ph}_{k_2}$) given almost any choice of k_1 and k_2,[6] but certain choices are better than others. Here, it will be useful for Ph_{k_1} to order the pitch classes by the generating interval, for reasons that will become clear shortly, so we choose $k_1 = 1$. For the other dimension, any value of k_2 will do, but since the function of this dimension is to split apart the two whole-tone collections, a good choice will be one that does so most effectively. The ideal choice from this perspective would be Ph_6, but Ph_6 (or $\mathrm{Ph}_{k/u}$ with $k = u/2$) is degenerate in a 12-tone universe: it can logically

[6] The exception is a choice such that k_1, k_2, and u are not mutually prime; see Yust (2015b). Ideally we should also apply some constraints on the choice of k_1 and k_2 relating to the compactness of regions, but determining exactly how this should be done would be somewhat involved and therefore beyond our present scope.

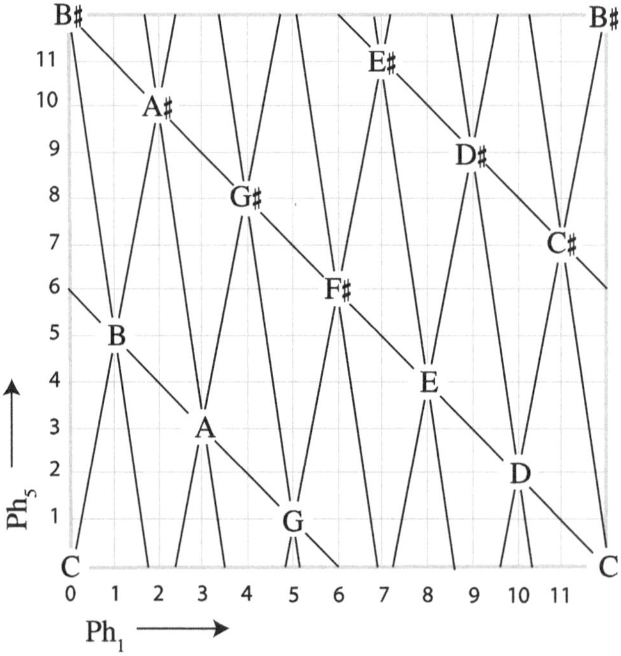

Figure 3. Geometrical *Tonnetz* of spelled chromatic trichords in $Ph_{1,5}$-space.

take only two values (0 and 6), so it is not really possible to ascribe any meaning to a Ph_6 path (except by "oversampling" to a larger universe). The next best choice is Ph_5, which represents the average circle-of-fifths position of a collection. Figure 3 shows a $Ph_{1,5}$ space triangulated by chromatic trichords. The simplicial Tonnetz embedded in this geometry has twice as many 2-simplexes than the one in Figure 2, two for each chromatic trichord, and it is topologically toroidal, with two cyclic dimensions rather than just one.

A simplicial Tonnetz *triangulates* a phase space, partitioning it into regions, such that each region is associated with a trichord (or more generally $(n+1)$-note chord) from the set of trichords belonging to the given *Tonnetz*. These regions correspond to all of the points belonging to the given trichord, in the sense that each possible weighting of the notes in the trichord corresponds to some point within the given region.[7] For a generated trichord, such as the chromatic trichords of Figure 3, there are actually two such regions, because there is an orientational ambiguity in one dimension. For instance, in $Ph_{1,5}$ space, the note F is an equal distance above or below the dyad {E,F♯}. If the trichord {E,F,F♯} is weighted more towards E than F♯, then it will fall in a region below the {E,F♯} dyad. If it is weighted more towards F♯, then it will fall in a region above. Alternately, we could imagine adding other notes to provide the context that distinguishes the two versions of the trichord. In the presence of, say, {C♯,D♯,G♯}, the chromatic trichord is {E,E♯,F♯}. In the context of {G,A,D} it is {E,F,F♯}.

Values of Ph_5, average circle-of-fifths positions, are of significant musical interest: Amiot (2017a) shows that the size of its associated Fourier coefficient, f_5, is the best measure

[7] Our drawings of the regions, with straight lines as boundaries, however, are not completely accurate and should be seen as an idealization. In fact, if the interval defining a given boundary has a larger magnitude in one dimension than the other, this imparts a sinusoidal curve to the boundary. In practical situations, the amount of curvature is typically slight and therefore can be safely disregarded.

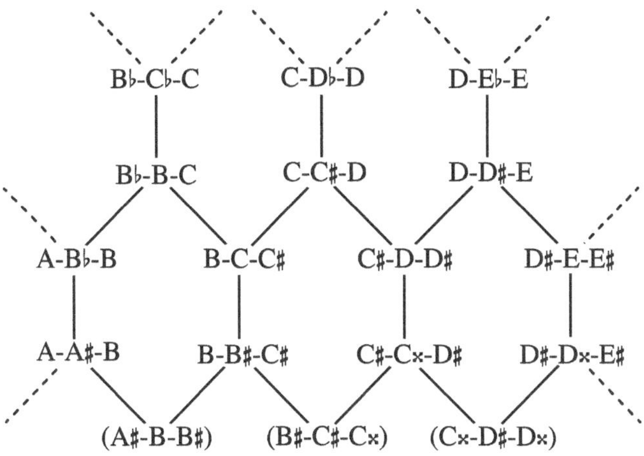

Figure 4. *Tonnetz* graph of spelled chromatic trichords.

of the *diatonicity* of pitch-class sets,[8] and it is overwhelmingly the most prominent harmonic quality of pitch-class distributions that occur across tonal music (Yust 2017b, 2019). For this reason, Ph_5 is the principal criterion for distributional key-finding algorithms and models of key perception (Cuddy and Badertscher 1987; Krumhansl 1990; Yust 2017b). Given this relationship to keys, Ph_5 paths associated with ic1 distinguish between diatonic semitones – the shorter way around the Ph_5 cycle and the one that occurs within a diatonic collection – and chromatic semitones, which go the longer way around the Ph_5 cycle and span beyond the limits of a single diatonic collection.[9]

In the *Tonnetz* of Figure 3, each chromatic trichord is made up of both kinds of semitone, one that corresponds to -5 in Ph_5 and one that corresponds to $+7$ in Ph_5, which sum to the regular whole tone: $(-1, -5) + (-1, 7) = (-2, 2)$. Because there are two ways to order the chromatic and diatonic semitones, there are two kinds of chromatic trichord. Therefore the *Tonnetz* graph corresponding to Figure 3 is not the same as the one corresponding to the simplicial complex in Figure 2; the *Tonnetz* graph for this space is the one shown in Figure 4, with 24 vertices and cyclic in two dimensions. Note that the sets at the bottom of the figure in parentheses are equivalent to the ones at the top: the identities of the pitch classes are still enharmonically identified. The distinctions are made only with respect to the intervals within the pitch-class sets. B♭-C♭-C is therefore distinct from B♭-B-C but equivalent to A♯-B-B♯. The graph has two kinds of cycles: one that goes through all of the pitch classes following the horizontal zig-zags (a cycle of 12 elements) and one that alternates swapping the semitone types (vertical edges) with left and right moves on the horizontal cycles (a cycle of 4 elements).

---

[8] Dmitri Tymoczko (personal communication, October 2018) has challenged this point with the argument that diatonicity is not an intuitive description of f_5 in other universes (say, for quarter-tones, $u = 24$). The obvious response to this is that f_5 has only been equated with diatonicity in contexts where 12-tET is taken for granted – in other words, "diatonicity" or Quinn's (2006) "diatonicness" has only ever referred to $f_{5/12}$, not $f_{5/u}$ for all u. Nonetheless, Tymoczko's challenge raises an interesting point, which is that certain concepts related to harmonic qualities should generalize away from a particular discretization of the octave. This is true of diatonicity: we have an intersubjectively robust intuitive sense of diatonicity in, say, 31-tone equal temperament. The appropriate generalization to capture this for $u > 12$ is $|f_{7/u}||f_{12/u}|/|f_0|^2$. That is, diatonicity is determined, generally, by the combined size of two components, f_7 and f_{12} (normalized by the total power, f_0^2). The special status of these particular components ultimately comes, presumably, from the fact that they give the best approximations to the acoustic perfect fifth, $\log_2(3/2)$, so an alternate, more basic, definition might be given by the correlation with the spectrum of the perfect fifth with f_0 (and values of k exceeding some threshold) excluded.

[9] For more on this distinction, see Yust (2015b, 2018b).

In the toroidal geometry, two kinds of chromatic trichord occupy different regions of the space, with distinct boundaries (overlapping in the shared whole-tone interval) but the same vertices. The same vertices can bound totally different regions of space because Ph_5 is cyclic. The differences between two trichords with the same pitch classes are contingent upon the meaning of Ph_5: to the extent that this is relevant, the distinction is relevant. This would be true given a tonal context: for instance in a context of A major, the chromatic trichord {E,E♯,F♯} has a different meaning (a move to the sharp side, tonicization of the relative minor) than {E,F,F♯} (a move to the flat side, mode mixture). For composers of the nineteenth century, orientations along the circle of fifths can have crucial musical meaning (see, e.g., analyses of Schubert's tonal plans in Yust 2015b, 2018b, 2020). In such contexts, spelling distinctions are used to indicate orientations around the circle of fifths, which is how they are commonly, though not exclusively, used in tonal music.[10] These Ph_5 distinctions might not be meaningful, in other contexts, such as in Webern's music. In that case we want a different kind of geometry, one that embeds the simplicial Tonnetz of Figure 2. We can derive such a geometry by folding a toroidal one like $Ph_{1,5}$-space.

2.3. *Mathematical definition of phase-space foldings*

In the two-dimensional, trichordal, case, defining a folding for a toroidal Tonnetz is easy to do given the appropriate conditions: a geometric *Tonnetz* in which triangles representing the same trichord overlap in one interval, and where the generating (duplicated) intervals project onto one axis without overlapping. The later condition can be met by the selection of the appropriate Ph_k for the given generator. In particular, for generator g/u (g and u coprime), the appropriate k is one that satisfies the equation $kg =_{\text{mod}(u)} \pm 1$. Through this relationship a phase space is uniquely associated with a specific interval – it will be the essential dimension for *Tonnetze* on sets of any cardinality generated by that interval.[11] For ic1-generated trichords, $Ph_k = Ph_1$, and for the ic5-generated trichords it is Ph_5. Or, for another example, the triad is a generated trichord in a mod-7 universe (Figure 5) and its generating interval ($g = 2$) projects neatly onto a Ph_3 axis ($2 \times 3 =_{\text{mod}7} -1$). Many important rhythms are also generated: the *tresillo* is a three-note rhythm generated by interval 3 in a minimal embedding universe of 8 (usually a dotted quarter in 4/4), and as a *Zeitnetz* in $Ph_{1,3}$ space (Figure 6) it has duplications. The duplicated rhythms are differently oriented in time, since Ph_1 is the dimension representing simple temporal proximity. This could be understood as two ways of grouping the same rhythm, 3-2-3 or 2-3-3, as illustrated on the right side of Figure 6. Such a distinction may be musically important in some contexts, in which case we would want to retain this toroidal version of the *Zeitnetz*. If it is not, however, we may want to fold this geometry to produce a simple cyclic *Zeitnetz*.

Let us refer to the axis that takes a projection of the generator cycle a *principal dimension* and the other as a *disambiguating dimension*, whose function is to separate two forms of the generating interval.

[10] Spelling itself is by definition a notational phenomenon, and as such it is accurately described by a system of voice leading on seven-note collections, or generalized key signatures (Tymoczko 2005). Although Ph_5 mimics the central axis of this voice-leading space when restricted to relatively even seven-note collections (Yust 2016), the two are derived from a very different set of assumptions and should not be conflated (Tymoczko and Yust 2019). However, it is routine in music to treat enharmonic notation as an imperfect tool for expressing an underlying musical reality, often relating to key or harmonic function (though by no means limited to that). To the extent that one believes key perception relates to pitch-class distributions – a theory that remains debatable, but is supported at this point by a large amount and variety of empirical evidence (Krumhansl and Cuddy 2010) – tonal composers' use of spelling often reflects an underlying reality expressed by Ph_5 relationships.

[11] The relationship of intervals to Fourier components (i.e. the index of a phase space, Ph_k) has been explored by Quinn (2006) and Amiot (2007). Both show that a one-to-one relationship does not hold for interval *content* per se, but does for intervals as generators.

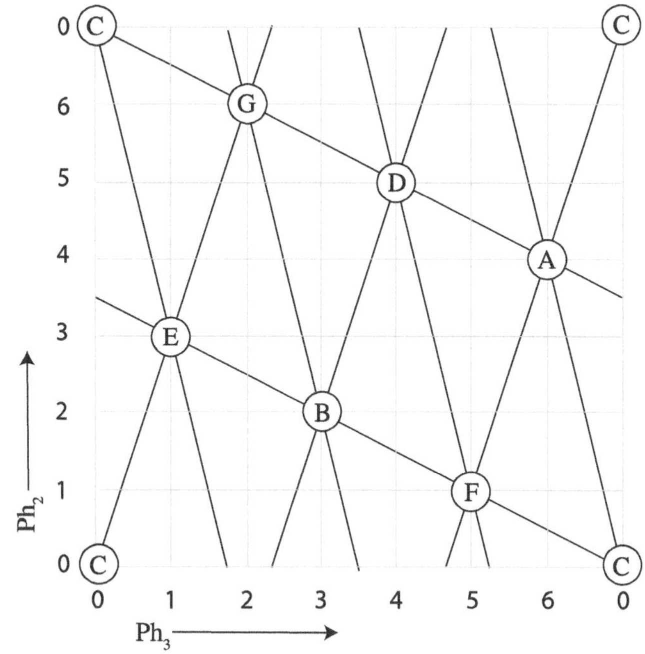

Figure 5. Geometrical *Tonnetz* of generic triads in $\mathrm{Ph}_{3,2/7}$-space.

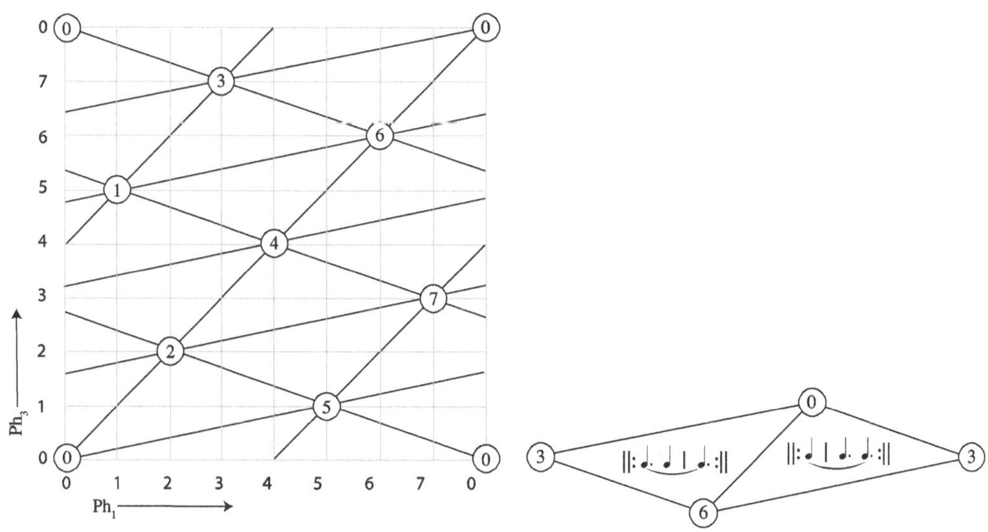

Figure 6. Geometrical *Tonnetz* of tresillo rhythms in $\mathrm{Ph}_{1,3/8}$-space.

Given a two-dimensional generated *Tonnetz* with axes x and y, let $f(x, y) = 0$ be an equation for the intervallic axis for the non-generating interval, which I will call a *boundary function*. If u is even (as in Figure 3 and Figure 6), then half of the pitch classes will satisfy $f(x, y) =_{\mathrm{mod}(u)} 0$ and the other half will satisfy $f(x, y) =_{\mathrm{mod}(u)} u/2$. If u is odd, as in Figure 5, all pitch classes will satisfy $f(x, y) =_{\mathrm{mod}(u)} 0$. For the (012) *Tonnetz*, we have $f(\mathrm{Ph}_1, \mathrm{Ph}_5) = \mathrm{Ph}_1 + \mathrm{Ph}_5$. For the

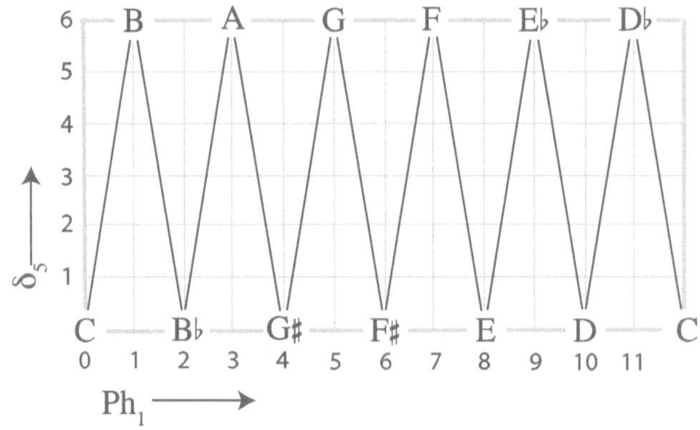

Figure 7. Folded chromatic trichord *Tonnetz*.

tresillo *Tonnetz* we have, similarly, $f(\mathrm{Ph}_1, \mathrm{Ph}_3) = \mathrm{Ph}_1 - \mathrm{Ph}_3$. For the generic triad *Tonnetz*, we have $f(\mathrm{Ph}_3, \mathrm{Ph}_2) = \mathrm{Ph}_3 + 2\mathrm{Ph}_2$. Note that the boundary functions necessarily pass through the origin and therefore have no constants.

In the case where u is even, we can reparameterize by retaining the principal dimension, x, and replacing the disambiguating dimension with $\delta = u/2 - |f(x, y)_{\mathrm{mod}(u)} - u/2|$, which gives the distance from the line given by the boundary function. The value of z is maximum where $f(x, y) = u/2$, and zero on the boundary function. The new space parameterized by x and δ equates the two forms of the generating interval. It has only one kind of cycle, the x cycle; in the δ dimension there are "mirror" boundaries, meaning that the image of a straight line (such as the one for the generating interval) "reflects" off the boundaries. It is useful also to add a subscript k to δ that recalls the original dimension $y = \mathrm{Ph}_k$. Figure 7 shows a folded space for the (012) *Tonnetz* in Figure 3, where $\delta_5 = 6 - |(\mathrm{Ph}_1 + \mathrm{Ph}_5)_{\mathrm{mod}12} - 6|$. The ic1 intervals are the image of a single straight line.

This satisfies our original goal: a well-defined geometry that embeds the simpler form of the generated trichordal *Tonnetz*, without duplications. However, the purpose of defining such a geometry, rather than simply using the simplicial form of the *Tonnetz*, is that it is potentially richer in meaning. The meaning of the cyclic dimension of the generating interval remains intact from the original phase space to the folded one, but the meaning of the newly defined δ parameter, and to what extent it inherits some meaning from the original phase space, demands further examination. The musical interpretation of the original space derives from the positions it assigns to other pitch-class sets and distributions, and the same is true of the folded space. Therefore understanding the meaning of the δ parameters involves consideration of how other kinds of pitch-class sets occupy the space.

An illustrative example, let us consider two versions of an (027)-*Tonnetz*. The optimum space for this *Tonnetz* (excluding the use of Ph_6) is the same as for chromatic trichords, $\mathrm{Ph}_{1,5}$. However, we might also be interested in thinking of (027)s as an element of $\mathrm{Ph}_{3,5}$ space. Krumhansl (1990) proposed this as a basic space of harmonies and keys in tonal music, and her space has been applied widely in music theory and music cognition research. I relate it to a number of issues in the analysis of tonal music in Yust (2015b). The Ph_3 parameter corresponds to a nearest even division of the octave into three parts, which can be understood as the three positions (root-third-fifth) of a triad, agnostic as to which position corresponds to the root. When coupled with Ph_5 and applied to primarily high-diatonicity distributions (as in tonal music), Ph_3 distinguishes dominant-side and subdominant-side harmonies with respect to a key (note that Ph_5 can then

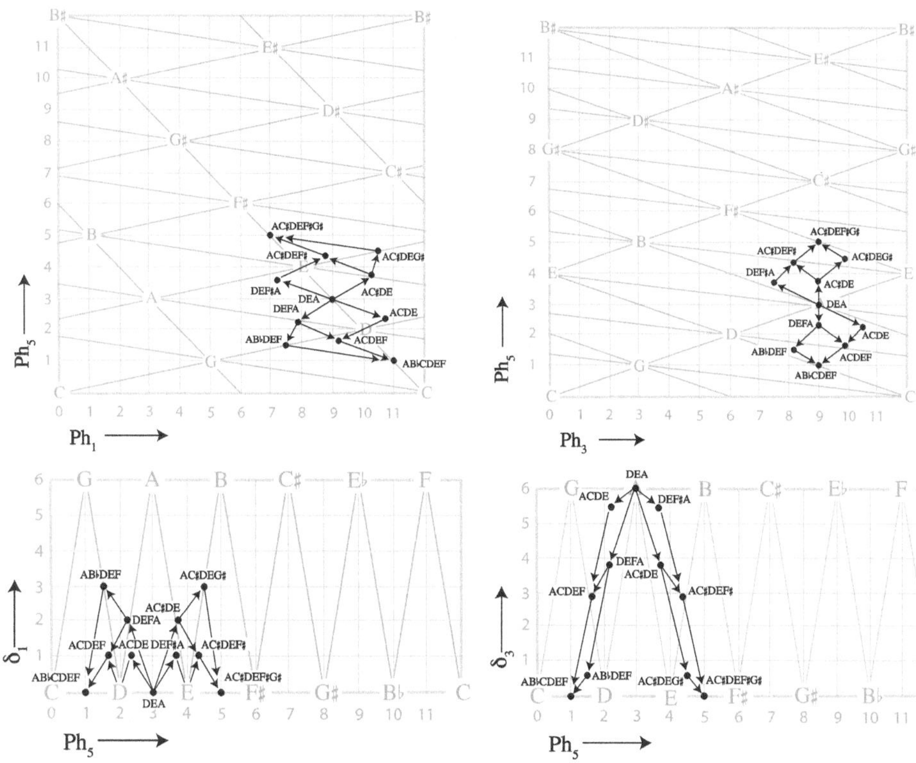

Figure 8. An (027)-*Tonnetz* in two spaces and foldings showing different resulting positions of major and minor triads.

determine which triadic position corresponds to the root).[12] I will therefore refer to this as the triadic space.

Figure 8 shows each of these (027) *Tonnetze*. In the $Ph_{1,5}$ space, two kinds of ic5 interval distinguish fourths from fifths. The boundary space (whole-tone axis) is defined by $Ph_1 + Ph_5 = 0$. In the $Ph_{3,5}$ space, the difference is one of triadic status. The shorter Ph_3 distance between notes a fourth apart corresponds to thinking of them as differing by one triadic position, as they normally would as the fifth and root of a major or minor triad. Taken the long way around, we understand the *fifth* to span one triadic position and the fourth to span two. For instance, in a context centered on a D major triad, the note C would be understood as a downward displacement of the root, and G an upward displacement of the third, so the fifth C-G spans one triadic position (root to third) while the fourth G-C spans two (third to root). The large Ph_3 distance of this interval reflects the relative unusualness of this situation. The boundary space is the whole-tone axis made by the sum of these two different kinds of fourth: $Ph_3 + 3Ph_5 = 0$.

The lower panels of Figure 8 show the result of folding each space to eliminate duplications of the (027) trichords. Considering just the locations of pitch classes and the *Tonnetz* itself, the two spaces look identical. However, if we consider the locations of various pitch-class sets or distributions, we see substantial differences, beginning with the (027)s themselves. In the δ_1 space, each (027) is located at the midpoint of its (02) dyad, since the (02) dyad determines the Ph_1 balance of the set. In the δ_3 space, on the contrary, the (027) is understood as a kind of triad, with both fourths representing potential triadic intervals, so it is located at the single

[12] See also Yust (2019).

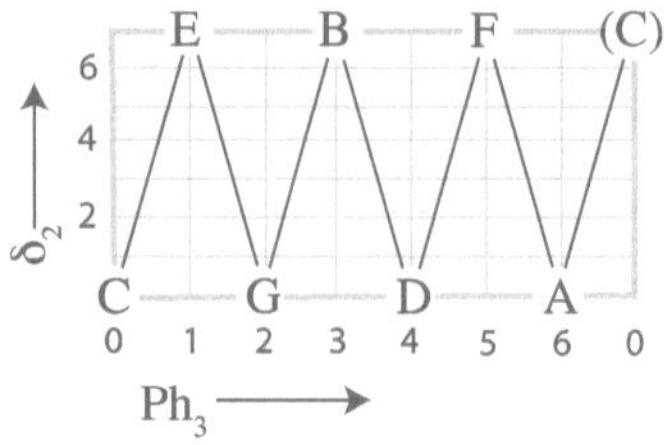

Figure 9. A folded generic-triad *Tonnetz*.

pitch class shared by both fourths. Figure 8 shows a series of pitch-class sets plotted in both the $Ph_{1,5}$ and $Ph_{3,5}$ spaces and the corresponding folded spaces, beginning from DEA and gradually adding, one at a time, pitch classes from the (027) a major third above (F♯, G♯, C♯) or below (B♭, C, F). The changes of Ph_5 are the same in both spaces (monotonically sharpward or flatward). But in δ_1 the sets remain anchored to the even whole-tone scale. In δ_3 they cross the space to the opposite side, reflecting the triadic significance of the added notes (which fill in a diatonic stack of thirds).

If u is odd, then we may define $\delta = u - |f(x,y)_{\text{mod}(2u)} - u|$. By taking $f(x,y)$ mod $2u$ rather than mod u we divide the line in half at the midpoint of its cycle from (0, 0) to (0, 0). The first half of the boundary line takes on a minimum value $z = 0$ and the second half the maximum $z = u$. For the generic triad *Tonnetz* in Figure 5, $\delta = 7 - |(Ph_3 + 2Ph_2)_{\text{mod}14} - 7|$, resulting in the space in Figure 9. The line from $(0,0)$ to $(0,u)$ is identified with (u,u) to $(u,0)$, resulting in a Möbius strip, a non-orientable space, which has been discussed, e.g., by Mazzola (2002).

2.4. *Higher-order foldings*

As we increase the cardinality of generated collections, we can define higher-dimensional *Tonnetze* in toroidal spaces by further differentiating forms of the generating interval. Then a similar method of folding may be applied to turn cyclic dimensions into bounded ones, until a single cyclic chain of intersecting n-simplexes remains, reflecting the basic *Tonnetz* graph. After the space is fully folded, it is homotopy-equivalent to the circle that acts as its principal dimension.

For example, a *Tonnetz* of chromatic tetrachords may be defined in $Ph_{1,4,5}$ space by differentiating one ic1 interval in the Ph_4 dimension and another in the Ph_5 dimension. The *Tonnetz* contains six copies of each (0123) tetrachord based on the possible permutations of the three forms of ic1, which correspond to tangent-space vectors $(-1,-4,-5)$, $(-1,-4,7)$, and $(-1,8,-5)$. The sum of these is the minor third interval $(-3,0,-3)$. By taking the cross product of one of the ic1s with the ic3, we get the equation of a plane for one of the (013)s shared by "enharmonically equivalent" tetrachords. There are three of these (one for each ic1 type): $(-1,-4,-5) \times (-3,0,-3) = (12,12,-12)$, $(-1,-4,7) \times (-3,0,-3) = (12,-24,-12)$, and $(-1,8,-5) \times (-3,0,-3) = (24,-12,-24)$. Reducing the products by a factor of 12, we get the equations of three boundary planes: $Ph_1 + Ph_4 - Ph_5 = 0$, $Ph_1 - 2Ph_4 - Ph_5 = 0$, and $2Ph_1 - Ph_4 - 2Ph_5 = 0$. By design only two of the three planes are linearly independent; the third is a sum of the other two: $(1,1,-1) + (1,-2,-1) = (2,-1,-2)$.

The folded space for this *Tonnetz* will retain Ph_1 as the cyclic dimension. The other dimensions (Ph_4, Ph_5) are replaced by barycentric coordinates λ_1, λ_2, and $\lambda_3 = 1 - \lambda_1 - \lambda_2$, defined for each cross-section of the space $Ph_1 = x$ by the points $(x,0,x)$, $(x,4,x+4)$, and $(x,8,x+8)$ where the three boundary planes intersect. These points define a triangular lattice of the cross-section with

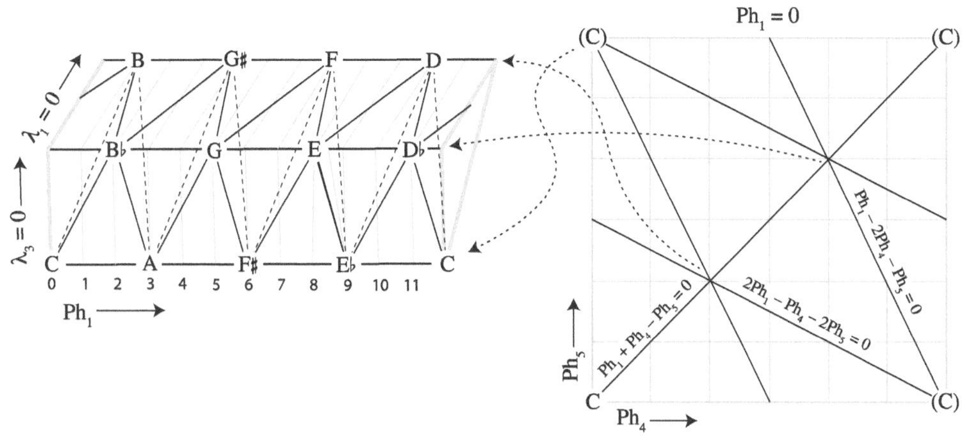

Figure 10. Folded (0123)-*Tonnetz* and a cross-section of the toroidal space it is derived from.

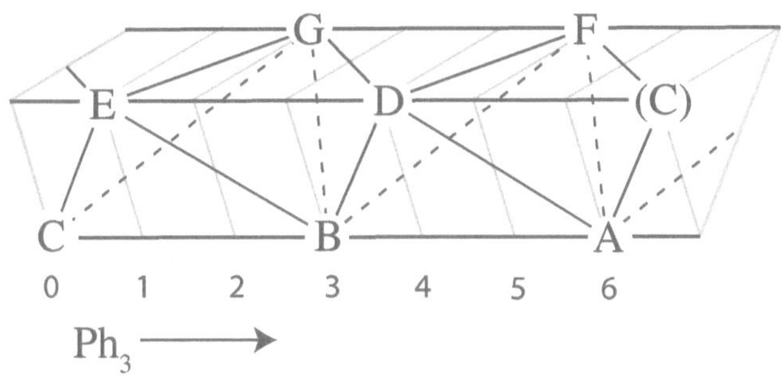

Figure 11. Folded generic seventh-chord *Tonnetz*.

six triangles, as shown on the right side of Figure 10, and the barycentric coordinates map all of these triangles onto one another.[13] The resulting space is illustrated on the left side of Figure 10.

For comparison, consider a *Tonnetz* of generic seventh chords generated by 2 mod 7 (a generic third). Starting with a $\mathrm{Ph}_{1,2,3}$ space, we can define three forms of the generating interval, $(-2,3,1)$, $(-2,-4,1)$, and $(5,3,1)$. Taking cross products with the sum $(1,2,3)$ (which represents the interval of the seventh), we have boundary planes $\mathrm{Ph}_1+\mathrm{Ph}_2-\mathrm{Ph}_3$, $-2\mathrm{Ph}_1+\mathrm{Ph}_2=0$, and $\mathrm{Ph}_1-2\mathrm{Ph}_2+\mathrm{Ph}_3=0$. The principal dimension is Ph_3, so for cross-section $\mathrm{Ph}_3=x$ barycentric coordinates can be defined on $(x/3, 2x/3, x)$, $((x+7)/3, (2x+14)/3, x)$, and $((x+14)/3, (2x+7)/3, x)$. The result is the *Tonnetz* shown in Figure 11. This geometry is similar to the one for the (0123) *Tonnetz*, but different in one important respect: because the universe is not divisible by three, there is a twist where the two ends of the triangular prism reconnect.[14] Furthermore, because the chord is so large relative to the universe ($n > u/2$), every note in the universe is connected to every other in this *Tonnetz*. Therefore, if it were treated merely as a graph, we could find a complete subgraph for every possible set in the universe, including every tetrachord. However, geometrically only the seventh-chord tetrahedra define a simplicial

[13] There are three barycentric coordinates given by projections onto the lines from each vertex of the triangle to the midpoint of the opposite edge, and their values are the proportion of the distance from the opposite edge. The three coordinates are constrained to sum to a constant, so only two of the three are independent.

[14] Compare to the trichord space of Callender, Quinn, and Tymoczko (2008) and Tymoczko (2011), which has the same topology. This space is derived differently, from voice-leading considerations, but is similar to the spaces discussed here in that it can be understood as a quotient of a direct product of three Ph_1 spaces.

decomposition. A different generating interval, such as a fifth, will make a complete cycle in the Ph_3 dimension of the space (e.g. C-G-D-A-(C)), so it will not actually outline a region, even though all of its notes are connected in one way or another. Thus, defining Ph_3 as the principal dimension is essential to making this the space for a generic seventh-chord *Tonnetz* (as opposed to, say, an (0134), diatonic stack of fourths, *Tonnetz*).

The use of barycentric coordinates can be generalized to simplices of any dimensionality, so this method of folding is generalizable to any one-interval generated collection. The entire process is:

(1) Define an n-dimensional phase space that includes one dimension with non-overlapping projections of the generating interval (i.e. Ph_k such that $gk =_{\mathrm{mod}(u)} \pm 1$).
(2) Define n unique tangent-space vectors that correspond to the generating interval, and are linearly independent.
(3) Take the cross product of each of these with the remainder interval. This defines n hyperplanes containing the intersections of two versions of the set.
(4) For each cross section, $Ph_k = x$, define barycentric coordinates using the n points where these hyperplanes intersect.
(5) The resulting space defined by Ph_k and the barycentric coordinates contains the desired non-redundant *Tonnetz* geometry.

This general scheme can be understood to capture the $n = 2$ case also, where the boundary hyperplanes are one-dimensional. These are split into one or two edges where the boundary function crosses each cross section, and the δs give distances from the endpoints of the edge, which can be understood as a trivial instance of barycentric coordinates.

2.5. *The heteromorphic diminished seventh chord*

The diminished seventh chord is a special kind of generated collection, a perfectly even collection where the remainder interval is equal to the generating interval. As such it represents a logical extreme of the kind of collections explored in the previous section: a *Tonnetz* of diminished seventh chords, considered just as basic *Tonnetz* graph or simplicial *Tonnetz*, is trivial, containing only one chord. As a musical object, though, a diminished seventh chord is potentially rich in multiple meanings, something that made it a an essential resource to eighteenth- and nineteenth-century composers. If the process of embedding a simplicial *Tonnetz* in a robust geometry is one of investing it with musical meaning, then even a chord that gives rise to a trivial *Tonnetz* might have a non-trivial network of meanings when embedded in different spaces.

The minimal embedding universe of a diminished seventh chord is $u = 4$ – i.e. the chord itself. Accordingly, its only available phase space is $Ph_{1/4}$, with $Ph_{2/4}$ being degenerate. Since three dimensions are required to make a four-note chord into a simplex, the procedure outlined in the previous section using phase spaces requires beginning from the direct product of three copies of $Ph_{1/4}$, which implies associating different meanings with each of these copies. Since we are interested in the diminished seventh chord as a component of tonal harmony, it makes sense to embed it in the non-minimal $u = 12$, in which case $Ph_{1/4}$ can be interpreted as $Ph_{1/12}$, $Ph_{3/12}$, $Ph_{5/12}$, all of which order the notes of the diminished seventh in the same way. We already have developed interpretations for each of these in the context of tonal harmony: $Ph_{1/12}$ is a position on the pitch-class circle, $Ph_{3/12}$ is a triadic location, and $Ph_{5/12}$ is position on the circle of fifths. I will use $Ph_{11/12}$ in place of $Ph_{1/12}$, however, so that interval values have their conventional interpretation (with 1 as an ascending semitone). The notes of the diminished seventh are equally spaced in all of these dimensions.

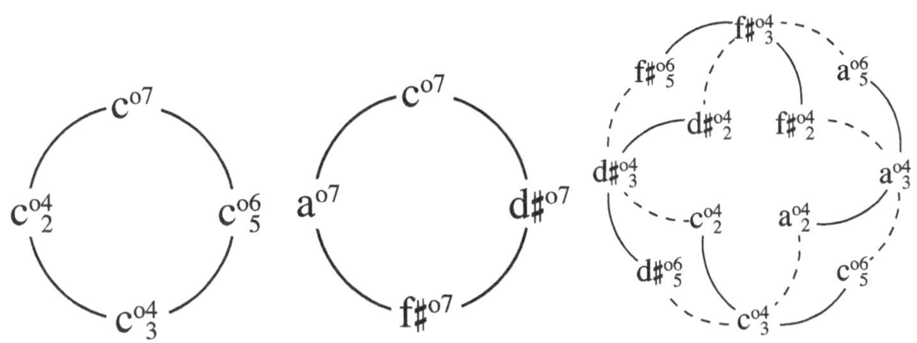

Figure 12. Graphs for folded (0369)-*Tonnetze*. On the right, dashed lines connect chords with the same bass, and solid lines chords with the same nominal root.

In each of these phase spaces, we can realize the generating interval of the diminished seventh either as an interval of length 3 (or −3) or −9 (resp. 9). In Ph_{11}, the distinction is between an ascending 3-semitone interval and a descending 9-semitone interval. This dimension can therefore represent differences of voicing or figured bass. The complete diminished seventh consists of three intervals of size 3 and one of size −9. The ordering of these will determine the position of the chord: 7, $^6_5$, $^4_3$, or $^4_2$. For instance, if we begin from C and fix the spelling of notes as C-E♭-G♭-B♭♭, then ordering the Ph_{11} interval values 3-3-3-(−9) produces a c$^{\circ 7}$, 3-3-(−9)-3 produces a c$^{\circ}\,^6_5$, and so on. The left panel of Figure 12 shows a *Tonnetz* graph for these different positions of the c$^{\circ 7}$.

Spelling can be used to distinguish between the two possible Ph_5 orientations of the 3-semitone interval, as a minor third (−3) or augmented second (9). This changes the figured bass along with the nominal root of the chord, so, with the Ph_{11} interval ordering fixed at 3-3-3-(−9) (from pitch class 0 to 9), the Ph_5 ordering (−3)-(−3)-(−3)-9 gives c$^{\circ 7}$, (−3)-(−3)-9-(−3) gives a$^{\circ}\,^6_5$, (−3)-9-(−3)-(−3) gives f♯$^{\circ}\,^4_3$, and 9-(−3)-(−3)-(−3) gives d♯$^{\circ}\,^4_2$. A corresponding *Tonnetz* graph is shown in the middle of Figure 12.

We can combine these two kinds of intervallic distinctions to make a *Tonnetz* of different possible nominal roots and positions, by defining the intervals of the seventh chord in $Ph_{11} \times Ph_5$ as $(3,9)$, $(-9,-3)$, and two copies of $(3,-3)$. For instance, $(3,9)$-$(-9,-3)$-$(3,-3)$-$(3,-3)$ is a d♯$^{\circ}\,^6_5$. By permutation of these intervals, 12 possible forms of the same diminished seventh are possible, and they make the *Tonnetz* graph shown on the right of Figure 12. This *Tonnetz* excludes the root position chords, because they have a different set of intervals: $(3,-3)$-$(3,-3)$-$(3,-3)$-$(-9,9)$. The root-position diminished seventh is an essentially one-dimensional structure because it has only two kinds of intervals.

The diminished seventh chord is perhaps the earliest instance where harmonic theory very clearly separated interpretations of the same set pitch classes as distinct harmonic objects. Bach (1949, p. 438) (orig. 1753), for instance, makes it clear that the idea of a diminished seventh as a common entity with distinct enharmonic interpretations was well established by the mid-eighteenth century. If we consider that the pitch-class distributions of tonal music very consistently maintain a strong diatonicity (high $|f_5|$), diminished sevenths, with their pitch classes equally spaced on the circle of fifths, were the earliest method that composers used to weaken f_5 and thereby create an ambiguity of diatonic position.

The distinctions made in Figure 12 might not be entirely sufficient to reflect the possible functions of a given diminished seventh, though. In particular, the notion of "nominal root" determined solely by Ph_5 orientation is somewhat unsatisfactory. For typical harmonies (triads and seventh chords), the concept of a root is associated with triadic distinctions, Ph_3 orientations,

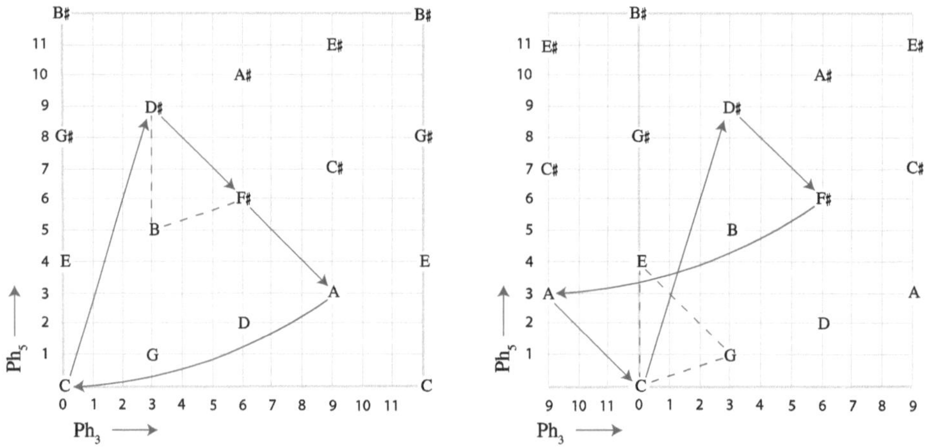

Figure 13. Different triadic orientations of the same diminished seventh chord, as a dominant (displacement of B major in an E minor context) or tonic-functioning, or common-tone, diminished seventh (displacement of C major).

as well as Ph_5 orientations. We have noted above that Ph_3 orientations divide the pitch classes up between three triadic positions (root, third, fifth, but without specifying which position corresponds to the root). A seventh chord, since it has four notes, must have two notes assigned to the same triadic position. With most seventh chords (dominant, minor, half-diminished), there is no ambiguity about the triadic orientation: the root and seventh are a step apart and belong to a single triadic position. Either the seventh is a displacement of the root or, in certain contexts, the root might be considered a displacement of the seventh (in which case the third might be called the "true root"). In a diminished seventh chord, since it is equally spaced, any two notes might be equally likely to be assigned to the same triadic position. Furthermore, there is not necessarily any reason why the Ph_5 orientation of notes (as sharpward or flatward, determining the spelling of the chord and therefore assignment of nominal root) should have to correlate in any particular way to their Ph_3 orientations – that is, the spelled root and seventh need not belong to the same triadic position, as they normally would in a diatonic seventh chord.

From the eighteenth century on, the diminished seventh has been conventionally understood as a displacement of a dominant seventh, or a dominant minor ninth with a missing root. This implies that the nominal fifth (as the seventh of the implied dominant) and nominal seventh (the minor ninth, or displacement of the root) belong to a single triadic position. This orientation in $Ph_3 \times Ph_5$ space is shown on the left of Figure 13, for a d♯$^{\circ 7}$ chord, whose displaced root would be B. This conventional interpretation can be generalized as a particular ordering of $Ph_3 \times Ph_5$ intervals: $(3,9)$-$(3,-3)$-$(3,-3)$-$(-9,-3)$. Here the notes interpreted as triadically stable, D♯-F♯ are also the furthest sharp. If we fix the spelling (Ph_5 orientation) but shift the Ph_3 orientation in the tonic direction (to the left), we get a different interval ordering, $(3,9)$-$(3,-3)$-$(-9,-3)$-$(3,-3)$, shown on the right of Figure 13. The furthest flat note (nominal seventh), C, is now triadically stable, while F♯ and A now share a triadic position and may be treated as displacements of G. The D♯, though triadically stable, is remote in Ph_5 and therefore can resolve to E to give a C major triad.[15] In other words, this $Ph_3 \times Ph_5$ orientation corresponds to the usual common-tone interpretation of the diminished seventh and shows why this distinct interpretation cannot be captured by spelling alone. Logically we can then extrapolate to a third possible interval ordering, $(3,9)$-$(-9,-3)$-$(3,-3)$-$(3,-3)$, in which the notes A and C are triadically stable and the

[15] Theoretically this reasoning could be inverted so that the chord resolves to G♯ minor with A and C resolving down by semitone. This is less compelling because of the general tonal tendency to associate descent in Ph_5 (decending fifths, ascending leading tones) with resolution.

sharp-side notes, D♯-F♯, are displacements of a hypothetical fifth, E. This inversion of the usual dominant-functioning interpretation might be understood as a minor version of the common-tone diminished seventh (a displacement of A minor).

The diminished seventh, as a 3-simplex, can be embedded in $\mathcal{T}^3$, thus potentially admitting of all three of these kinds of distinctions simultaneously. Specifically, we can make a *Tonnetz* on the diminished seventh {C,E,F♯,A} whose intervals in $\text{Ph}_{11}\times\text{Ph}_3\times\text{Ph}_5$ are $(3,3,9)$, $(3,-9,-3)$, $(-9,3,-3)$ and $(3,3,-3)$. The 24 possible permutations of these intervals partition the $\text{Ph}_{11,3,5}$ space into 24 tetrahedral regions, all with *the same four vertices*, differing only in their orientation in the space. Even though they are, in a sense, all versions of the same chord, they make a non-trivial *Tonnetz*, with certain interpretations overlapping in shared tetrachords and others being more distant.[16]

We may then simplify this space through various types of folding, which remove the intervallic distinctions (and thereby some of the types of diminished seventh) and the cyclic homology of one of the dimensions. If we forget the triadic (Ph_3) distinctions, we get a *Tonnetz* of 12 regions for the graph on the right side of Figure 12. If we instead forget the distinctions relating to direction around the pitch-class circle (Ph_{11}), between 3 semitones in one direction and 9 in the other, then we have a *Tonnetz* of 12 triadic/diatonic forms of a diminished seventh (3 functions by 4 spellings), whose graph is isomorphic to the one for the 12 distinctions of spelling and position. In this latter space (cyclic in Ph_3 and Ph_5 and folded in Ph_{11}), the diminished seventh chord lives in a universe of interval classes, where it is not possible to orient along the pitch-class circle. However, spellings and triadic distinctions remain in effect, so sharp-flat orientation and functional orientation are possible. Also, recall from above that even after a dimension is folded, it still effects the underlying topology, so that proximity in Ph_{11} continues to effect proximity of points within the regions (insofar as these points represent pitch-class distributions), even as the global sense of orientation in Ph_{11} has been lost.

We can also go one step further, folding two dimensions of the large diminished-seventh *Tonnetz*, to get the simple cycles of interpretations on the left of Figure 12. This is precisely analogous to the folded chromatic tetrachord and diatonic seventh-chord *Tonnetze* of the previous section, differing only in the cardinality of the resulting cyclic *Tonnetz* graph (with 4 elements instead of 12 or 7). However, the diminished seventh chord is different from these in that its remainder interval is equivalent to its generating interval, which makes it possible to have a *Tonnetz* in which *all three dimensions are folded.* This fully simplified diminished seventh *Tonnetz*, with no interval duplications, is in a bounded tetrahedral space with the pitch classes of the diminished seventh as vertices, all of the intervals between them (minor thirds and tritones) as edges, and the four diminished triads as faces bounding the entire space. This can be derived directly from the full $\text{Ph}_{11,3,5}$ space by defining barycentric coordinates based on the distance of each point from the nearest plane for each diminished triad, mapping the 24 tetrahedral regions defined by the diminished seventh onto one another. Pitch classes not in the *Tonnetz* fall on the face opposite the pitch class two semitones away. So, for instance, pitch classes D and F♯ are mapped onto one another and fall on the (E♭, F♯, A) face (opposite C). The resulting set of pitch classes on a common face therefore, interestingly, forms an (01478), which is special as the only example of a simple pitch-class set in $u = 12$ that is perfectly balanced ($|f_1| = |f_5| = 0$) but not transpositionally symmetrical (Amiot 2017b; Milne, Bulger, and Herff 2017).

[16] Note that the total is 24 (rather than $4^3 = 64$) because certain intervals are excluded, namely those that are long in two dimensions at once. We already noted that the diminished seventh interval has this property (it is long in both Ph_1 and Ph_5). Similarly the long Ph_3 interval (the minor third that does not span triadic positions) is never the augmented second (indeed, it would hard to interpret the tonal meaning such an interval), and also is never allowed to be the major sixth interval.

3. Two-interval generated *Tonnetz*

Thus far we have only considered *Tonnetze* on one-interval-generated collections, but the principles of embedding a *Tonnetz* in a toroidal phase space and folding dimensions to equate duplicated intervals can be applied more widely. Here we consider another class that includes most of the other examples of theoretical interest, collections generated by two distinct interval types. The primary difference is that instead of a *Tonnetz* on a single transposition-type, we will now combine multiple transposition-types in a single *Tonnetz* or *Zeitnetz*, including, but not limited to, inversionally related transposition-types (e.g. major and minor triads).

The simplest case of two-interval generated *Tonnetze* is the trichordal *Tonnetze*, such as the standard triadic *Tonnetz*. Any trichord that is not one-interval generated can be generated by two distinct intervals *a*-*b*, and will have a remainder $(-a - b)_{mod(u)}$. Its inversion will then be given by taking these in the opposite order, *b*-*a*. For instance, the standard triadic *Tonnetz* results from setting $a = 4$ and $b = 3$, giving major triads, 4-3-5, and minor triads, 3-4-5, where $5 = (-a - b)$ is the remainder. Note, however, that this method of generating the standard *Tonnetz* is not unique. We could also use, e.g., $a = 5$ and $b = 4$, giving major triads in "second inversion," 5-4-3, and minor triads in "first inversion," 4-5-3.

Trichordal *Tonnetze* generated by two intervals are not foldable in the way described in the previous section, because all of their intervals are unique.[17] Larger two-interval generated *Tonnetze*, however, are. One example is a seventh-chord *Tonnetz* described by Douthett (1997) and Yust (2018a). This extends the standard triadic *Tonnetz* with another minor third. To get the transposition types for this *Tonnetz*, we list all unique permutations of the intervals 3, 3, and 4, and append the remainder $12 - a - b - c = 2$. The result is three types of seventh chords: dominant sevenths, 4-3-3-2; minor sevenths, 3-4-3-2; and half-diminished sevenths, 3-3-4-2. Note that inversionally related transposition types will always be included. Nonetheless we still count them separately, since some transposition types will have inversional symmetries (like the minor seventh). The folding for this seventh-chord *Tonnetz* equates the two minor thirds, so that there are no duplications of the minor seventh chords. The resulting topology may be understood as a fattened two-dimensional torus – that is, it has two cyclic dimensions and one bounded dimension. Each boundary contains a whole-tone collection with a (026) trichordal *Tonnetz*. These (026) trichords are connected to points on the opposite boundary to complete dominant or half-diminished seventh chords. Minor seventh-chord simplexes then also result from these connections, filling in the space between the 3-3-4-2 and 4-3-3-2 tetrahedra to give a complete simplicial decomposition of the space.

We can construct other two-interval generated tetrachordal *Tonnetze* similarly, and they will have one bounded dimension (the disambiguating dimension) and two cyclic dimensions (principal dimensions), which results in either the same topology as the seventh-chord *Tonnetz* (an orientable fattened two-dimensional torus) or a similar non-orientable topology. An example of the latter type, also mentioned in Yust 2018a, is a space of scalar tetrachords generated by 1, 2, and 2 – that is, tetrachords in the older sense. The tetrachords are the "phrygian" 1-2-2-7, "dorian" 2-1-2-7, and "ionian" 2-2-1-7. The remainder is coprime to 12, so the second method of folding needs to be used in this case, resulting in a non-orientable space with a single boundary. This boundary houses a trichordal (015)-*Tonnetz*. Connecting the (015)s across the bounded dimension to the note that fills in the major third makes phrygian and ionian tetrachords, and the dorian tetrachords fill in the space between these.

Since two-interval generated *Tonnetze* always have two cyclic dimensions remaining after folding, they can be effectively visualized by projection onto a two-dimensional torus. For the

---

[17] However, this is not the only possible method of folding. See the section on spherical *Tonnetze* below.

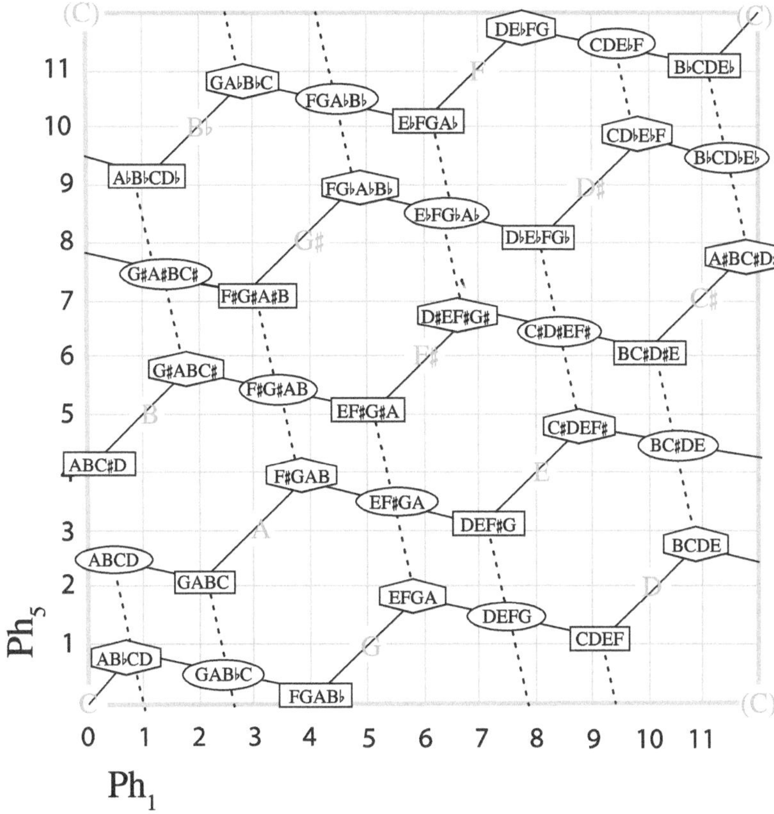

Figure 14. *Tonnetz* of scalar tetrachords, 1-2-2-7, 2-1-2-7, and 2-2-1-7, projected onto $Ph_{1,5}$-space. Dashed lines show "voice-leading"-type transformations that hold the remainder interval (the bounding fourth) constant. Solid lines show scalar extensions that remove the lowest note and add a note a fifth above or vice versa.

scalar-tetrachord *Tonnetz*, for example, the best choice of cyclic dimensions (which can be determined by maximizing Fourier coefficients) are Ph_1 and Ph_5. In Figure 14, the tetrahedra are projected into $Ph_{1,5}$-space by taking the Fourier phase values of their tetrachords (i.e. taking circular averages of their pitch classes in each dimension). The $Ph_1 \times Ph_5$ space may be compared to Noll (2018) *degree space* and corresponding concept of width-height duality, or height versus tone character (Clampitt and Noll 2011). Scalewise tetrachords embody this duality in the sense that they are equally compact in Ph_1 and Ph_5.

Although I have advocated throughout for the importance of an underlying geometry, including dimensions that may be rendered homologically trivial through folding, throughout this section and the next two, we will operate at a somewhat more general level, remaining agnostic about the dimension that has been folded and working instead with the *Tonnetz* graph and the topologically significant dimensions of the phase-space geometry (the ones that define the cyclic components of the homology group). In the (0135)/(0235) *Tonnetz*, the topologically significant dimensions are Ph_1 and Ph_5. Projection into this two-dimensional space eliminates the unspecified third dimension.

Two-interval generated *Tonnetze*, with 3 and 4 (in $\mathbf{Z}_{12}$) as the two intervals, formalize the common musical notion of tertian harmony. Such constructions generalize the standard *Tonnetz* to higher cardinality by considering the generating thirds as an essential feature, and increasing the cardinality, which corresponds to dimensionality in the topological construction. The Douthett seventh-chord *Tonnetz* is one example. A more exotic set of seventh chords would result from using two major thirds, as opposed two minor thirds (augmented-major sevenths, 4-4-3-1; major sevenths, 4-3-4-1; and minor-major sevenths, 3-4-4-1). We can also extend further to cardinality 5, or ninth chords. As the cardinality increases, we may also increase the number of foldings, which means the number of cyclic dimensions remains constant at two. All such *Tonnetze* will house a standard triadic *Tonnetz*, whose original form can be recovered by projecting it onto the cyclic dimensions.

The two ninth-chord *Tonnetze* of greatest musical interest are those generated by 3, 3, 3, and 4, and by 3, 3, 4, and 4. The former is a *minor ninth-chord Tonnetz* with remainder 11 and the latter is a *major ninth-chord Tonnetz* with remainder 10. The foldings are somewhat different in that the first identifies three intervals (the minor thirds) and the second identifies two pairs of intervals. Therefore the minor-ninths *Tonnetz* is somewhat simpler, including only four chord types. Figure 15 shows its projection in $Ph_{4,5}$-space. The edges of the *Tonnetz* graph are also shown, with dotted lines for voice-leading-type relationships, meaning that they hold the position of the ninth interval constant while moving one other note by semitone. Solid lines show the extensions of the chain of thirds, where the initial note in the chain is removed and another note added at the end, or vice versa. The major-ninths *Tonnetz*, with six chord types, has an optimum projection in $Ph_{3,5}$ space, as shown in Figure 16. Even this projection is not perfect, though, because the two chord types fall in the same location, the dominant ninth (4-3-3-4-10) and the minor-major ninth (3-4-4-3-10 – i.e. a minor triad with a major seventh and major ninth). To show the *Tonnetz* graph, then, doubled dotted lines indicate two distinct connections to the co-located dominant and minor-major ninths, while the solid lines to and from these show individual

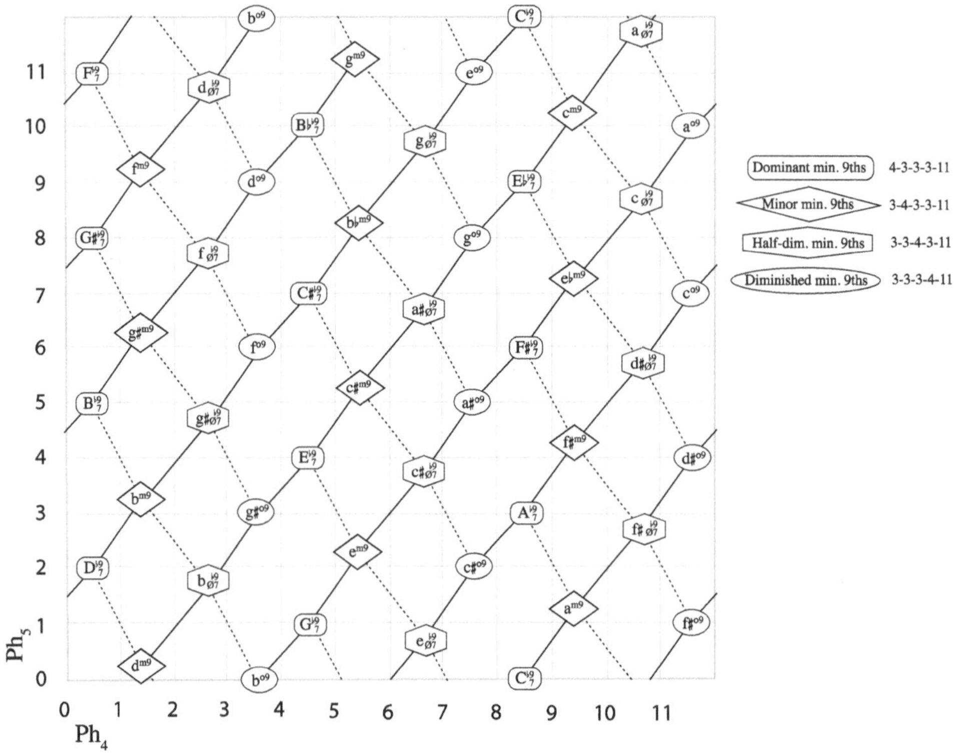

Figure 15. *Tonnetz* of minor ninth chords projected onto $Ph_{4,5}$-space. Dashed lines show "voice-leading"-type connections, which keep the remainder interval in place. Solid lines show extensions of the chain of thirds.

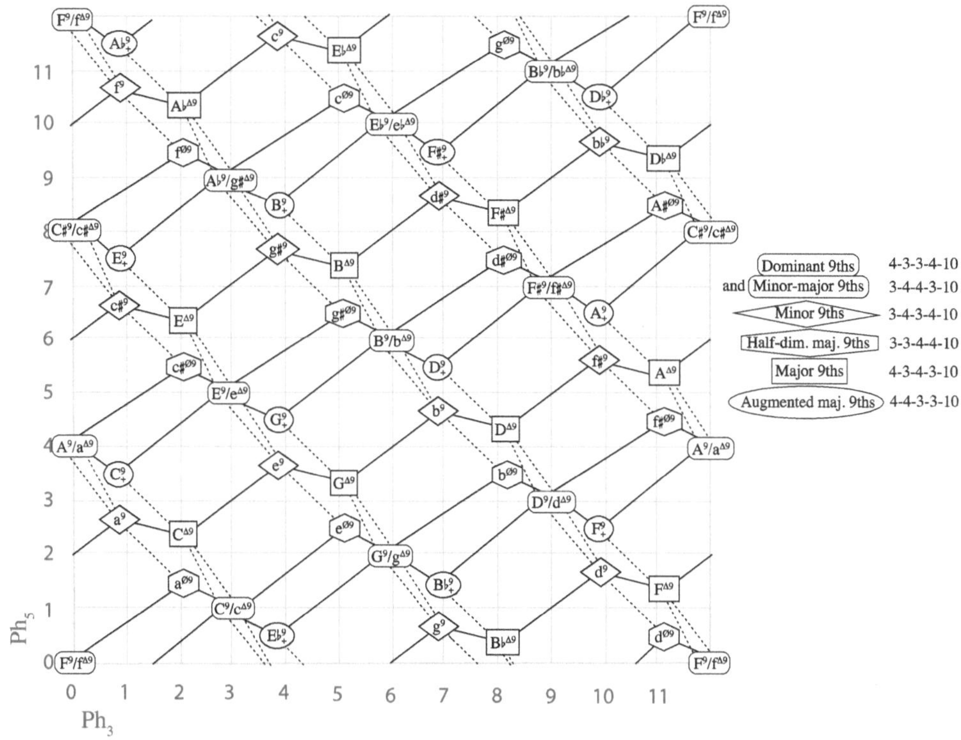

Figure 16. *Tonnetz* of major ninth chords projected onto $Ph_{4,5}$-space. Dominant ninths and minor–major ninths are represented by the same points in the space, and double lines show independent connections to each of these. Dashed lines show "voice-leading"-type connections, which keep the remainder interval in place. Solid lines show extensions of the chain of thirds.

connections (from left to right, augmented ninth → dominant ninth → half-diminished ninth, or half-diminished ninth → minor-major ninth → augmented ninth).

The minor-ninths *Tonnetz* has an interesting relationship to the diminished-seventh *Tonnetze* discussed in Section 2.5, because diminished sevenths are one of the shared subsets (between dominant-minor and diminished-minor ninth chords). The minor ninths *Tonnetze* at different levels of folding therefore all have diminished-seventh sub-*Tonnetze*. In the fully ramified 5-toroidal version of the *Tonnetz*, the sub-*Tonnetz* is the full 4-toroidal diminished seventh *Tonnetz*. When folded down to two cyclic dimensions, as in Figure 15, we see that the diminished-seventh sub-*Tonnetz* retains one cyclic dimension, here shown as Ph_5 distinctions, so that each diminished seventh chord corresponds to four vertically aligned edges in the network, at $Ph_4 = 0$, 4, and 8. The historical origins of diminished-seventh enharmonicism are therefore reflected in this *Tonnetz*: originally conceived as a dominant minor ninth chord with an elided root, it is required to retain a distinct enharmonic identity, as reflected in the dimensionality of the folded minor-ninths *Tonnetz*.

We can make an interesting *Zeitnetz* that is very similar to the major-9ths *Tonnetz*, generated by 3 and 4, but with $u = 16$, giving a remainder of 2 rather than 10. This *Zeitnetz* includes a number of interesting timelines that are used in some Latin American and West African musical genres, the best known being the Son clave of Afro-Cuban music (or Kpanlogo timeline of Ghana – see Agawu 2003, 157–161): 3-3-4-2-4 (a rotation of 4-3-3-4-2). This *Tonnetz* also includes other timelines discussed by Toussaint (2013), the Rhumba timeline (3-4-3-2-4, a rotation of 4-3-4-3-2) and the Gahu or Highlife timeline (3-3-4-4-2), as well as the inversions (i.e. retrogrades) of these, and one that he refers to as the "rap timeline" (4-3-2-3-4, a rotation of 3-4-4-3-2). The

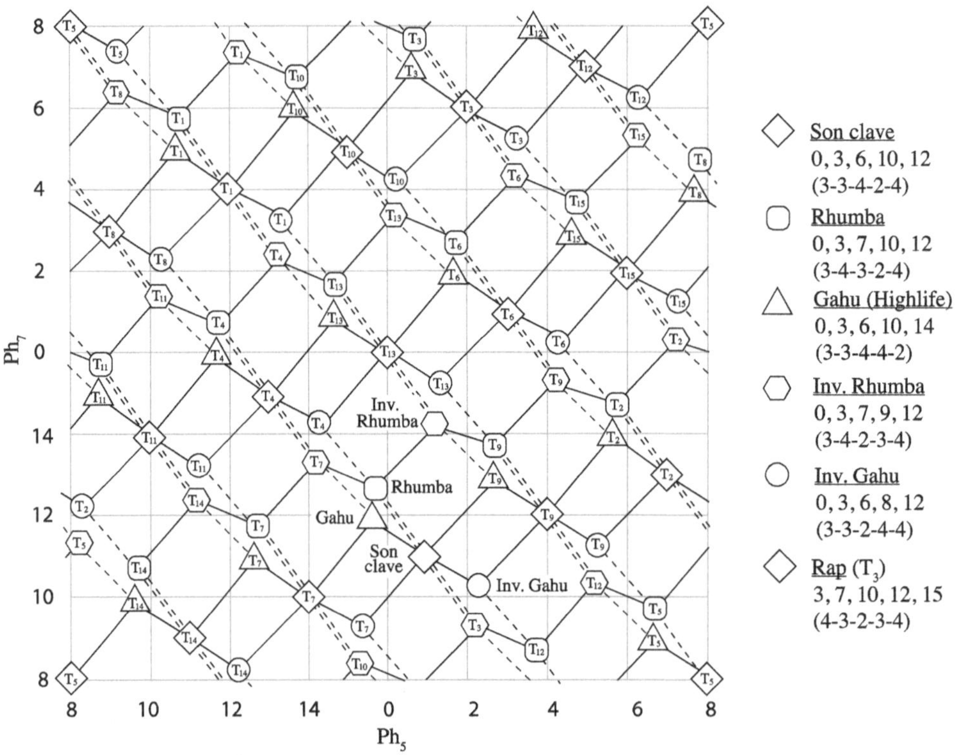

Figure 17. *Zeitnetz* generated by 3, 3, 4, and 4 in $u = 16$ projected in $Ph_{5,7/16}$-space. Son clave and the rap timeline occupy the same positions in the space, and double lines show independent connections to each of these. Dashed lines show "voice-leading"-type connections, which keep the remainder interval in place.

best projection of this *Zeitnetz* is in $Ph_{5,7/16}$-space, which much like the major-ninths *Tonnetz*, superposes two rhythms ("rap" and Son clave), as shown in Figure 17. The rap timeline is therefore given in T_3, the rotation that is superposed on the standard rotation of the Son clave. This phase space represents the rhythms as approximations to maximally even 5-in-16 (interval pattern 3-3-3-3-4) rhythms and subsets of maximally even 7-in-16 (interval pattern 3-2-2-3-2-2-2) rhythms.

4. Uniform *Tonnetze*

Another type of two-interval generated *Zeitnetz* can be constructed on what Osborn (2014) calls *Euclidean rhythms*, following Gómez-Martín, Taslakian, and Toussaint (2009) and Toussaint (2013), which include many rhythms common in popular music. These are defined as the beat-class sets that can be generated by permuting the adjacency intervals of a maximally even rhythm.[18] Although maximally even sets are generated when n and u are coprime (Clough and Douthett 1991), if the generator is not an adjacency interval, then distinct collections can result from permuting these.

[18] Toussaint's definition is different than Osborn's, effectively only including the maximally even patterns (which, it turns out, can be generated though a version of the Euclidean algorithm). Osborn's definition is somewhat preferable since a new term is not needed for maximally even patterns.

For present purposes, what is significant about Euclidean rhythms is that they can be understood as two-interval generated patterns where the *remainder interval is equal to one of the generating intervals*. Because the intervals need not relate to the Euclidean algorithm in the manner described by Toussaint, I will call these *uniform Tonnetze*. In uniform *Tonnetze* as compared to other two-interval generated *Tonnetze*, an additional folding is possible, so that it can be reduced to one cyclic dimension.

The simplest non-trivial example of such a *Tonnetz* is generated by 2 and 1 mod 6, which could be realized as quarter notes and eighth notes in 3/4, or as steps and thirds in a whole-tone scale. The latter realization gives a *Tonnetz* consisting of (0248) and (0268) tetrachords in 12-tET. This is folded according to the process described above, resulting in a space with only one cyclic dimension remaining, as it would be for a *Tonnetz* on generated tetrachords. The boundaries of the space are two augmented-triad Möbius strips. Figure 18 gives its graph, projected in $Ph_{2/6}$ but with the two set types separated. Note the lack of connection between sets that share a common augmented triad. The augmented triads occur on a boundary where two of the major-third intervals (two whole-tone steps) are equated, but one, the sum of the two whole-tone intervals of the set, remains distinct. Therefore there are actually three geometrically distinct forms of each augmented triad.

A similar *Tonnetz* generated by the intervals of the maximally even pentatonic scale, 2, 2, 2, 3, 3, contains two chord types, pentatonic scales and dominant major ninth chords, as shown in Figure 19. The choice of generating intervals differing by a minimal amount (1) guarantees, for a uniform *Tonnetz*, that one of the chord types will be maximally even. Assuming that the cardinality n is coprime to u, the maximally even collection is generated and the generating interval will project onto the axis Ph_n or $Ph_{(u-n)}$ without overlapping. In other words, to use a folded phase space for this kind of *Tonnetz*, the dimension that we retain as a cyclic dimension should be the one indexed by the cardinality of the collection. This is Ph_5 for the pentatonic *Tonnetz*. The form of the graph is determined by the number of intervals, two of one type and three of the other (as opposed to two of both types, as in Figure 18). For instance, a *Tonnetz* generated by 2, 2, 1, 1, 1 (pentatonic subsets and ninth-chord subsets in a mod-7 universe) would

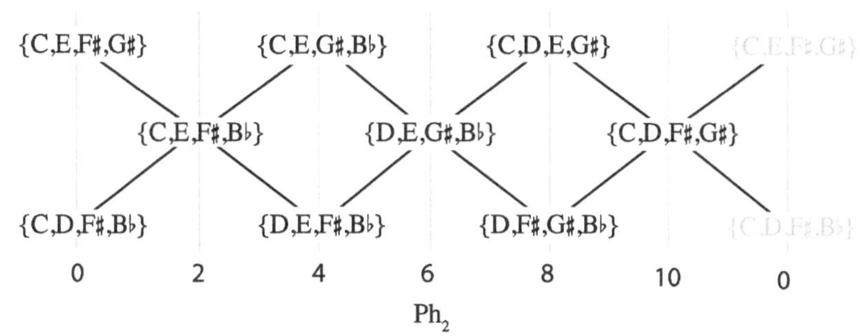

Figure 18. Graph for a uniform *Tonnetz* generated by 2, 2, 4, 4.

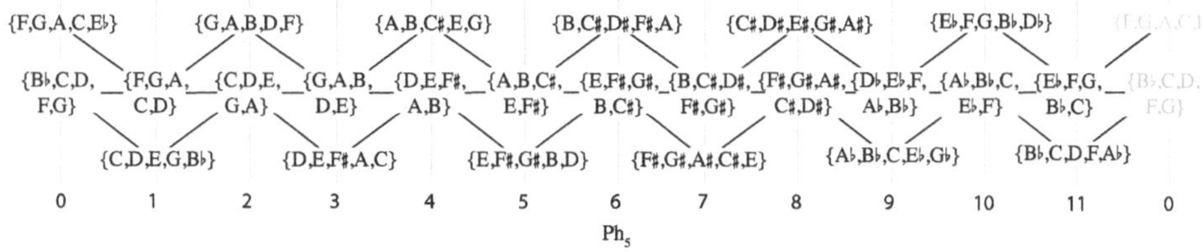

Figure 19. Graph for a uniform *Tonnetz* of pentatonic scales and dominant ninths generated by 2, 2, 2, 3, 3.

look like Figure 19, only cycling after 7 of each chord type instead of 12. Similarly, a *Zeitnetz* generated by 2, 2, 2, 1, 1 would have a central cycle of eight rotations of a *cinquillo* rhythm (2-1-2-1-2) with detours through rotations of 2-2-2-1-1.

In the previous section, we distinguished two kinds of links in two-interval generated *Tonnetze*, those of voice-leading type, where the remainder interval is fixed, and others that extend the chain of generating intervals in one direction or the other. In the uniform *Tonnetze*, this distinction breaks down, because the remainder interval is not distinguishable from a generating interval. Therefore, in cases like these, where the generating intervals differ by 1, the *Tonnetz* connections given by simple permutations are precisely the set of single-semitone voice leadings. Networks like this that include all and only the single-semitone voice leadings are discussed by Tymoczko (2011). If the pentatonic sets in Figure 19 are replaced by their complements, the result is a network of diatonic and acoustic scales, an important voice-leading network for Tymoczko.

This coincidence should be surprising, because the two kinds of network are derived from very different theoretical premises. But note that they are equivalent only as chord-based networks, not as geometrical *Tonnetze*. Section 2 above explained how Tymoczko (2012) creates a geometrical *Tonnetz* out of chord-based voice-leading networks by defining a dual space. For the pentatonic example, we can take as our objects hyperplanes in the five-dimensional voice-leading space. The possible rotations and translations of these hyperplanes define a dual five-dimensional space, where certain points represent pitch classes in the sense of representing a hyperplane in voice-leading space that holds a certain pitch class constant. This is precisely analogous to the two-dimensional (dyadic) case. The important point is that in all cases, the dual space has a cyclic dimension for translation that corresponds to the cyclic transpositional dimension of the voice-leading space which corresponds to the *directed voice-leading sum* (Cohn 1998). The *Tonnetz* graph or the simplicial *Tonnetz* can therefore be geometrically embedded in voice-leading space with no essential change to Figure 19 except for changing the meaning of the x-axis from Ph_5 to transposition of a five-note collection. It turns out that, for any size of collection n, these two quantities, Ph_n and the projection onto the central axis of the n-dimensional voice-leading space, are close approximations of one another as long as we restrict our attention to relatively even chord types. This is informally presented in Yust (2015b) and addressed in greater mathematical detail in Tymoczko and Yust (2019). Therefore, these two kinds of geometry will embed the same simplicial *Tonnetze*, with the same topology – the significant differences between them only emerge if we consider what kind of musical objects occupy the space between the pitch classes.

Another interesting example of a uniform *Zeitnetz* can be built from the maximally even 6-in-16 beat-class set 3-3-2-3-3-2. This rhythm may be understood as a repeated tresillo (3-3-2). Permutation of these intervals gives distinct beat-class sets 3-3-3-2-3-2 or 3-3-3-3-2-2, or some rotation of these. Biamonte (2014) calls the latter, a common rhythm in rock, ragtime, and other popular music genres, the "double tresillo" (see Cohn 2016). The resulting *Tonnetz* can be folded into one cyclic dimension, giving the graph shown in Figure 20.

Higher-dimensional uniform *Tonnetz* of scale types are also possible. The 7-in-12 maximally even scale is the diatonic, 1-2-2-1-2-2-2, and permutation gives the acoustic scale or melodic minor, 1-2-1-2-2-2-2, and the "whole-tone plus one" scale 1-1-2-2-2-2-2. The *Tonnetz* graph in Figure 21 is similar to that for the pentatonic *Tonnetz*, but with more scale types. The cyclic dimension is Ph_5 and a spine of connected diatonic collections tours the cycle with short detours to acoustic collections and long detours to whole-tone-plus-one collections. (Note, however, that, as with the other graphs in this section, only the horizontal positions labeled with the cyclic axis reflect the *Tonnetz* geometry. The vertical positions are chosen for convenience to separate collections that project onto the same Ph_5 location.)

An interesting feature of maximally even patterns like the diatonic, however, is that they can be defined as two-interval generated patterns through multiple distinct pairs of generators. For

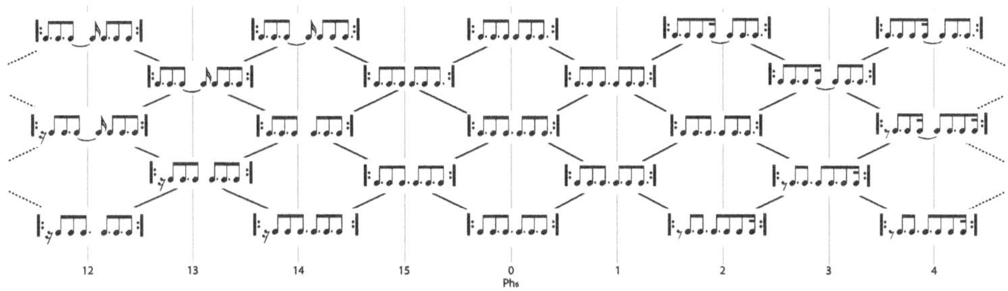

Figure 20. Portion of a graph for a uniform *Zeitnetz* generated by 3, 3, 3, 3, 2, 2. In the middle row are the maximally even repeated tresillos (3-3-2-3-3-2), the top and bottom rows have rotations of the double tresillo (3-3-3-3-2-2), and between these are the other pattern, 3-3-3-2-3-2. The graph cycles after going through the eight distinct rotations of the repeated tresillo, of which five are shown.

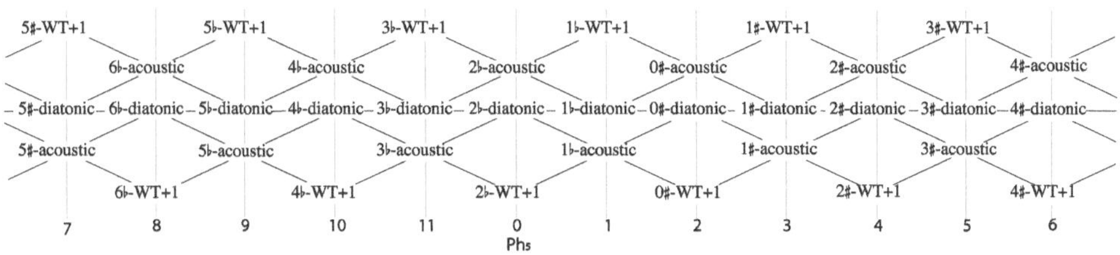

Figure 21. The graph of a uniform *Tonnetz* of scales generated by 2, 2, 2, 2, 2, 1, 1. The scales are labeled using Hook's (2011) system for spelled heptachords, which counts the number of sharps and flats needed to spell the collection with seven distinct letter names. For instance {G♯, A, B♭, C, D, E, F♯} is the 1♯ whole-tone-plus-one scale, because it is spelled with two sharps and one flat, and the difference then is one sharp.

example, we could treat the thirds as generators of the diatonic, constructing a scalar *Tonnetz* with a tertian logic. As Meredith (2011) observes, this captures what are typically taught as the standard tonal scales, including the harmonic minors and majors (4-4-3-4-3-3-3) as well as the melodic minor/acoustic (4-4-3-3-4-3-3), but not the whole-tone-plus-one. In addition, this *Tonnetz* includes a more exotic type of collection including a diminished second (i.e. enharmonic variant of a single pitch class), 4-4-4-3-3-3-3, e.g. C-D♯-E♭-F♯-G-A-B.

5. Spherical *Tonnetze* and other topologies

Each of the n-dimensional *Tonnetz* geometries constructed above includes a collection of hyperplanes with $(n-1)$-dimensional *Tonnetze* that make up the boundaries of the regions of the n-dimensional *Tonnetz*, corresponding to the subsets of the set classes involved. For instance, regions of the (0258)/(0358) *Tonnetz* are bounded by planes with (025), (026), (036), and (037) *Tonnetze*. In the n-torus (unfolded) *Tonnetz*, these sub-*Tonnetze* are $(n-1)$-toruses, and they retain this topology when a single dimension of the n-torus is folded.[19] With higher-order foldings, the homologies of the sub-*Tonnetze* are simplified in the same way as the parent *Tonnetz*. For instance, the uniform pentatonic *Tonnetz* of Figure 19 is a network of 4-simplexes (pentatopes) folded to have only one cyclic dimension (represented by Ph_5). The sub-*Tonnetze* are tetrahedral, with (0246) and (0258) *Tonnetze* making the boundary of the whole space, (0247) making the

[19] Although they may be reduced or mapped onto one another. For instance, in the 3-torus version of the seventh-chord *Tonnetz*, there are two distinct planes with (025) *Tonnetze* and two with (037) *Tonnetz*. These are mapped onto one another in the folded version. Each (036) *Tonnetz* has eight triangles as a plane in the 3-torus, but only four in the fattened 2-torus.

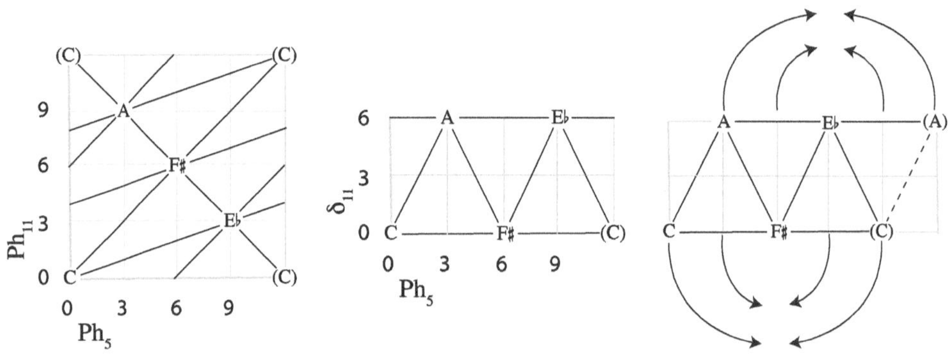

Figure 22. (036)-*Tonnetze* at different levels of folding (toroidal, cyclic, and spherical).

boundary between the two pentachord types, and (0257) as the boundary within the central cycle of pentatonic scales. Each of these tetrahedral *Tonnetz* are also cyclic in one dimension.

What happens, then, in the extreme case discussed in Section 2.5, of a totally bounded *Tonnetz* on a perfectly even collection? Figure 22 shows the process undergone by an (036) sub-*Tonnetz* at different levels of folding of the parent diminished seventh *Tonnetz*. It begins as a 2-toroidal *Tonnetz* and remains so with one dimension folded (i.e. when the diminished sevenths contain three different types of minor thirds based on distinctions in two dimensions, such as $Ph_{1,5}$ or $Ph_{3,5}$). With two foldings, when the diminished seventh *Tonnetz* becomes a cyclic network of four tetrahedral regions, the diminished triad *Tonnetz* is also reduced to a simple cyclic network of triangles bounded by tritone axes, as shown in the middle panel of Figure 22. In this (036) *Tonnetz* there are no longer distinct types of minor third, but there are two kinds of tritone. For instance, if Ph_5 is the cyclic dimension, the sum of two minor thirds is a diminished fifth, but the remainder of two minor thirds is an augmented fourth. In other words, all of the (036)s share two common tones, but those that share a common ic3 are adjacent in the *Tonnetz*, whereas those that share a tritone are not. When the parent diminished seventh *Tonnetz* is reduced to its simplest form, as a single tetrahedral space with no duplications, these duplicated tritones are also equated, as shown on the right side of Figure 22, resulting in an (036) *Tonnetz* that has, as the surface of a tetrahedron, a spherical topology. This means that it is a compact surface with a trivial fundamental group. In other words, this (036) *Tonnetz* is the logical extreme of removing the homological elements of a toroidal *Tonnetz*.

A spherical surface can be illustrated in stereographic projection, as on the left of Figure 23, which places one pole in the center, the opposite pole in a circle around the perimeter (i.e. this circle actually corresponds to a single point), and shows longitude lines as radii. We can see from this how transpositions by minor third and tritone map onto rigid symmetries of the sphere. The graph of the spherical (036) *Tonnetz*, on the right of Figure 23, is different than the graph of the cyclic and toroidal (036) *Tonnetze*. While all duplications of (036) trichords are already removed from the cyclic version (the one in the middle of Figure 22), the *Tonnetz* graph adds additional edges linking the tritone-related (036)s. The result, then, is a complete graph on four vertices (the graph-theoretic analog of a 3-simplex).

The construction of the tetrahedral *Tonnetz* can be generalized to higher-dimensional simplices, creating a family of *n*-dimensional hyperspherical *Tonnetze*. However, as a trichordal, 2-dimensional, *Tonnetz*, the construction is unique to the (036) set class (or, equivalently, a ♩♩𝅗𝅥 rhythm), because it is the only generated trichord with the remainder interval of a tritone, which can be equated with its inversion. Or, we might observe that the particular symmetries of the sphere used to satisfy the transposability criterion are order-4 and order-2, so the only available

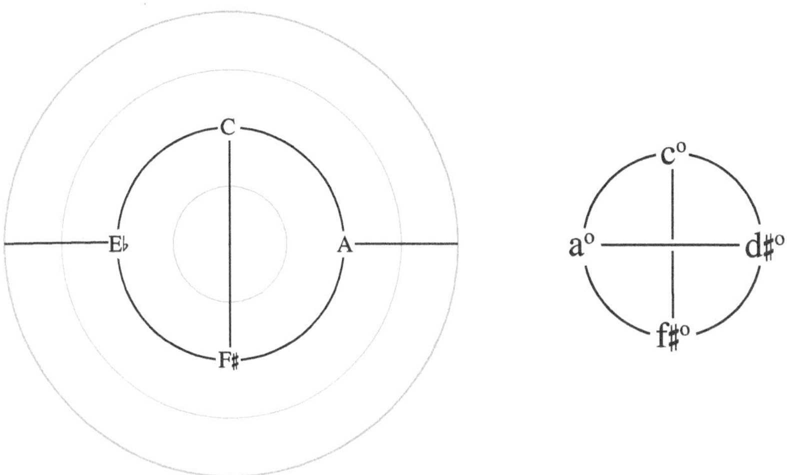

Figure 23. The tetrahedral *Tonnetz* in stereographic projection and its *Tonnetz* graph.

intervals are $8^{ve}/4$ and $8^{ve}/2$. However, there is one other way to construct a spherical trichordal *Tonnetz* with a different symmetry group.

One difference between the spherical topology and all the other *Tonnetze* we have constructed by folding n-dimensional phase spaces is its Euler characteristic, χ, which is an important topological invariant. This is defined as $\chi := V - E + F$, where V is the number of vertices in the simplicial complex, E the number of edges, and F the number of faces. The Euler characteristic is a topological property shared by the simplicial *Tonnetz* and the geometric *Tonnetz*. For all torii, $\chi = 0$. The folding process defined in Section 2 preserves χ, because it removes an equal number of edges and faces. But the one illustrated in Figure 22 removes two edges without removing any faces. This simplicial complex therefore has the Euler characteristic of a sphere, which is $\chi = 2$. To remove an edge by folding without changing the vertices, we need a cycle with exactly two vertices on it – i.e. a tritone axis. To get $\chi = 2$ we need exactly two of these, which occurs only in the (036) or equivalent *Tonnetz*.

However, there is one other way to change χ with V fixed, and that is to add faces without adding edges. Consider the (024) whole-tone *Tonnetz* in Figure 24. After being folded to a cyclic trichordal *Tonnetz*, it has a major-third boundary, which has exactly three vertices and three edges. Understood as a *graph*, then, it has a 3-clique (triangle) corresponding to each augmented triad. Geometrically, however, the three edges are colinear, so they do not actually define a region in the space. If such a region did exist, as Figure 24 shows, the *Tonnetz* geometry would be spherical.[20] Nothing in the phase space corresponds to this region, though – that is, we can see that such a geometry, with an octahedral symmetry group, could exist in principle, but we do not yet have a musical interpretation of it.

We find a related phenomenon, however, which does have musical significance, if we return to voice leading as the basis of our geometry, specifically trichordal voice-leading *Tonnetz* where u is divisible by 3, such as the standard triadic *Tonnetz*. Cohn (1997) initially linked the standard triadic *Tonnetz* to the concept of voice leading while formalizing it as a network of common-tone relationships, yet Tymoczko (2012) ultimately demonstrated the full repercussions of the

[20] Reasoning from χ can thus show that these are the only two examples of two-dimensional spherical *Tonnetze*. We could also observe that the existence of a spherical *Tonnetz* satisfying the transposability condition implies the existence of a Platonic solid with u vertices and whose symmetry group has a cyclic subgroup of order u. By process of elimination we are left then with only these two, tetrahedral and octahedral. For example, there is no cubic *Tonnetz* because the order-8 subgroups of the symmetry group of a cube are not cyclic.

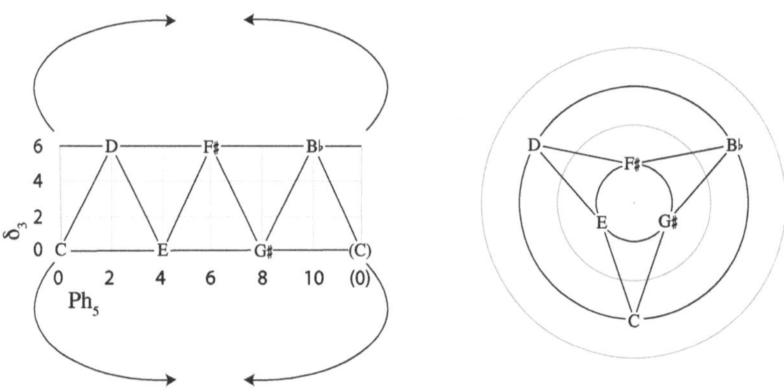

Figure 24. Derivation of the octahedral *Tonnetz* from a cyclic *Tonnetz*.

voice-leading conception of the *Tonnetz* using voice-leading geometries. Recall from Section 2 that geometrical *Tonnetze* can be constructed in dual voice-leading spaces. In the trichordal case, points in the dual space correspond to planes in the voice-leading space, while points in the voice-leading space correspond to planes in the dual space. The dual space therefore has a point for each pitch class (corresponding to the voice-leading plane that holds that pitch class constant), and each set of three pitch classes in the dual space defines a plane, which corresponds to a point in the voice-leading space belonging to the given trichord. Motions along that dual-space plane correspond to rotations about the point for that trichord in voice-leading space. We therefore have a dual-space plane for each major and minor triad, and can define triangular regions on each of those planes bounded by the trichord intervals.[21] Rotations of these planes around shared dyads ("flips") then correspond to edges in the chord-based network (a *Tonnetz* graph) in voice-leading space. Therefore, piecing together all of these regions we have a two-dimensional subspace of the three-dimensional dual voice-leading space corresponding to the traditional *Tonnetz*. Technically this subspace is not a surface because it intersects itself in multiple places, and for the same reason it does not properly satisfy the definition of a *Tonnetz* geometry given in Section 1. This situation is analogous to that of the dyadic voice leading *Tonnetz* in Section 2.1.

This voice-leading version of the traditional *Tonnetz* gives a distinct geometrical meaning to the same *Tonnetz* graph and simplicial *Tonnetz* as the phase-space version described in Section 3. The equivalent *Tonnetz* graphs result from the same general correspondence between Ph_n and the n-note voice-leading space that we found in uniform *Tonnetze* for maximally even collections in Section 4: for relatively even chords, Ph_n approximates a projection onto the central axis of the n-note voice-leading space. As a result, the same basic *Tonnetz* graph can be geometrically realized in two distinct ways.

However, because major and minor triads are *not* maximally even, the voice-leading space we have constructed for the traditional *Tonnetz* is not a satisfactory, or at least an incomplete, representation of triadic voice leading. As Tymoczko (2011, 2012) points out, the traditional *Tonnetz* inaccurately represents two-semitone voice leadings like {CEG}→{CFA♭} or {CEG}→{C♯EG♯} as relatively large as compared to the two-semitone "relative" voice leading,

[21] This can be done by specifying that the balanced voice leading plane occurs within the region, which is then defined as a convex hull around that point. Note that the balanced voice leading is the same plane for three major-third related triads, so these regions all cross one another in the space. For instance, the C major region and E major region cross in a line that corresponds to balanced voice leadings that hold E constant, C major and A♭ major cross in a line of balanced voice leadings holding C constant, and all three planes intersect in a point that corresponds to balanced voice leadings between trichords of sum class 11 ($= 0+4+7 = 11+4+8 = 10+5+8 \bmod 12$). The illustration in Tymoczko 2012 redraws the *Tonnetz* to eliminate these visually confusing intersections, which is helpful for visualizing its structure, but somewhat obscures the underlying geometry.

e.g. {CEG}→{CEA}. This is because a voice leading like {CEG}→{CEA} passes through an augmented triad {CEG♯} which has not been included in the network. Including these as a necessary intermediary between all two-semitone voice leadings between major and minor triads gives a network of single-semitone voice leadings that accurately represents distances in the voice-leading space. Adding augmented triads to the network is similar to doing so in the octahedral *Tonnetz* of Figure 24 above: each augmented triad adds a face to the space without adding any new vertices or edges. Therefore, the Euler characteristic of the standard triadic *Tonnetz*, $\chi = 0$, increases by 4 when considered as a proper voice-leading *Tonnetz*, even without accounting for the many intersections of triadic regions. According to the classification theorem of 2-manifolds, there is no way to consider such a simplicial complex as belonging to a surface (a connected compact 2-manifold), because the maximum Euler characteristic of a surface is that of a sphere, $\chi = 2$.

Tymoczko's *Tonnetz* therefore raises an interesting question about topology and musical meaning: what is the significance of a *Tonnetz* being embeddable in a manifold? A manifold is defined as a topology that is locally Euclidean – specifically, in a 2-manifold, every point has a region that is homeomorphic to the Euclidean plane. The triadic voice-leading *Tonnetz* fails this criterion in multiple places: at the major third intervals, which border three triadic regions, and on the intersections within triadic regions that share a common pitch-class sum. The locally Euclidean feature might be likened to reading a map: while it may be impossible to assign consistent orientations globally (for instance, on the surface of the earth), if we restrict our attention to a local region, we can imagine that the dimensions are fixed and extend infinitely. In the standard *Tonnetz*, this imagined Euclidean space is what Harrison (2002) called the "unconformed" *Tonnetz*. This is clearly how Riemann thought of the *Tonnetz* (Gollin 2011). As it was originally conceived, as a map of just tuning relationships, it is in fact a noncompact Euclidean space, but Riemann applied it in contexts where no such literal tuning differences were present. Rather the Euclidean space is imagined, which is possible to do on the toroidal *Tonnetz* as long as we maintain that only a limited extent of the space is "visible" from a given point at one time. Clearly analysts following Riemann have found this locally Euclidean aspect of the *Tonnetz* to be of great value, particularly when talking about tonal harmony from a perceptual standpoint. When Rings (2011), for instance, talks about Riemann's "intentional paths" he explicitly models the perception of tonal harmony as something locally Euclidean. In fact, the qualities of brightness-darkness and dominantness-subdominantness Rings discusses track closely with the dimensions of the $Ph_{3,5}$ space that can embed the standard *Tonnetz* and is posited as a perceptual space for tonal harmony by Krumhansl (1990). Cohn (2012, 64–67) explicitly advocates for the standard *Tonnetz* on the grounds of its locally Euclidean properties, even while maintaining that he is primarily interested in voice leading and that the *Tonnetz* does not accurately model this, for exactly the reasons that Tymoczko points out.

What Tymoczko shows us, then, is that when using the *Tonnetz* in this way, we cannot also imagine that it is a map of voice-leading relationships, at least not in the strict sense of voice leading modeled by voice-leading geometries. We can think of it as embedded within a two-dimensional phase space like $Ph_{3,5}$ with the structure of a manifold, but it is important then to recognize chords as instances of pitch-class distributions, a different sort of musical object than the chords of voice-leading spaces. Indeed, this does seem to be native to, e.g., Cohn's thinking, when he says,

> There are advantages to maintaining individual tones as primary objects, rather than prepackaging them into triads. The *Tonnetz* can track continuities among individual pitch classes, yielding analytical information that would otherwise be difficult to recover, and provides locations for other pitch-class combinations that arise in nineteenth-century music. These include consonant dyads and dissonant seventh chords that would otherwise need to be referred to a consonant triad via expansion or contraction, forcing interpretations that might be underdetermined or even arbitrary. (Cohn 2012, 67)

Cohn, that is, views triads as existing in a space that mixes chords of different cardinality, which is precisely what a space of pitch-class distributions does. One can make limited claims about voice leading in such a space, grounded in the mathematical association of Ph_n with the central axis of n-note chord space that holds for relatively even chords, but it is important to recognize that this does not amount to a robust theory of voice leading in Tymoczko's sense, and that if we want to talk about voice leading in this way, we cannot have a *Tonnetz* with the structure of a manifold.

Analogous points can be made about voice leading between seventh chords. As mentioned in Section 3 above, Douthett (1997) constructs a *Tonnetz* of dominant, half-diminished, and minor seventh chords, generated by 3, 3, 4 (remainder 2). When folded to equate the minor thirds, the boundaries of this *Tonnetz* are two whole-tone planes containing the (026) subsets of the dominant and half-diminished chords. At this stage, the seventh-chord *Tonnetz* has a toroidal topology, with a bounded third dimension. However, it still has duplications of the tritones, which can be identified to create a spherical sub-*Tonnetz*, as illustrated in Figure 22.[22] When the tritones are identified, the folded phase space now has fully connected (0268) tetrachords ("French sixth chords"), and to continue to think of it as a 3-manifold, we would have to understand these (0268) simplices as regions (adding six faces to compensate for the six edges), which they are not in the original phase space. As a folded phase space, this seventh chord *Tonnetz* now only has one non-trivial homological element, a cyclic homology associated with Ph_4, oblique to the spherical (036)-*Tonnetze*, which is also the phase-space dimension that mimics a voice-leading axis for four-note chords. This folded phase space also has fully connected diminished seventh chords. They do not define regions, however, just as the augmented triads of the toroidal triadic *Tonnetz* are fully connected but do not define regions. If were to treat them as such, we would obtain a *Tonnetz* graph equivalent to Tymoczko's (2011) seventh-chord voice-leading lattice. As in other cases, this voice-leading lattice can be realized geometrically in the dual voice-leading space. The seventh-chord regions of this voice-leading *Tonnetz* intersect where they share pitch-class sums, analogous to the two- and three-dimensional cases, and have singularities at (036) faces, which belong to three regions (dominant, half-diminished, and diminished). The diminished seventh chords are essential to accurately reflecting voice-leading distances because they are intermediaries in progressions like G^7-$b^{\varnothing 7}$, analogous to triadic relatives like C-a.

6. Topology of musical concepts

Tonnetz graphs, as defined at the outset of this article, are easy to produce and admit of many possibilities, and a variety of kinds of *Tonnetze* and *Zeitnetze* have been illustrated here. Using phase spaces, such networks can be embedded in musical geometries, and we have found that these geometries relate to a number of significant existing musical concepts. Due to the special role of simplicial complexes in the study of topology, we have been able to readily apply this method to revealing some of the topology of musical concepts, such as the cyclic homology of diatonicity and harmonic function. More complex topologies can create musical heteromorphs, distinct meanings assigned to identical pitch-class intervals, but our geometries can also forget

[22] A three-dimensional toroidal *Tonnetz* was also proposed by Gollin (1998), but, lacking the minor seventh chords, it is impossible to think of Gollin's *Tonnetz* as actually filling any space, including a three-dimensional toroidal one. Tymoczko (2012) points out that Gollin did not really explicitly realize the geometry of his *Tonnetz*, and therefore he might have been thinking of something like a voice-leading *Tonnetz*, in which case the tritone duplications in his illustration would be superfluous. If, on the other hand, he had something like Douthett's *Tonnetz* in mind, as he suggests in referring to the geometry as toroidal, the tritone duplications are necessary, and we could give a retroactive theoretical justification for them along the lines suggested in Yust (2018a), using phase spaces. Nevertheless, Tymocko's critique illustrates the importance of establishing clear theoretical foundations when proposing something like a geometry of chord relations.

these differences, altering the global homology of the space but without loosing the local topology, the common-tone based concept of proximity and the local directionality of the parent phase spaces.

A notable example of a particularly heteromorphic musical object is the fully diminished seventh chord. Through variability in its orientation in pitch space, the circle of fifths, and functional space, it can potentially support a surprisingly large *Tonnetz* based entirely on changes of interpretation of a single chord. Such a *Tonnetz* might be useful in rationalizing the many ways diminished sevenths can be reinterpreted in nineteenth century music, not only enharmonic changes but also changes of function, such as common-tone function versus its standard dominant function, and more distant types of reinterpretation can be distinguished from less distant ones. The historical importance of the diminished seventh chord in giving rise to richly ramified heteromorphic play in chromatic harmony may be explained, in fact, by the role of $Ph_{3,5}$-space as a basic syntactical space of keys and harmonic functions. As a perfect representative of the f_4 dimension of Fourier space, the diminished seventh is completely oblique to the $Ph_{3,5}$ plane, and therefore has a unique ability to shape-shift within tonality.

It is worth reflecting here that a 12-dimensional vector space on pitch-class distributions can be defined directly, with each dimension corresponding to the weighting of one pitch class. Because the Fourier transform is actually a change of basis, this is actually the same as the larger Fourier space. What the Fourier transform allows us to do is to project this larger space into one of fewer dimensions, such as our phase spaces, which forget, in particular, differences between pitch-class sets that relate specifically to chord type. Phases spaces are therefore not in a one-to-one relation with pitch-class distributions, but we can define canonical paths between points (such as simple interpolation, or "pitch-class cross-fade") that work in all phase spaces.

Phase spaces are not the only way to impart geometrical meaning to the abstract simplicial networks of generalized *Tonnetze*, but they are particularly well suited to this task because, as a space of tonal distributions, proximity in Fourier space derives directly from common-tone content. Because of this, they offer a particularly general way of geometrically embedding possible *Tonnetze*. The other prominent way of imparting musically meaningful geometry to a *Tonnetz* is through voice-leading geometry. While this is also in principle somewhat generalizable, the desire for network distances to reliably reflect voice-leading distances gives rise to a more limited set of possibilities. Where the two give distinct musical meaning to the same networks, we have a particularly interesting point of contact between theories based on very different basic principles.

Acknowledgements

Thanks to Dmitri Tymoczko, whose comments on an earlier version were instrumental in clarifying and improving the content and presentation of this work.

Disclosure statement

No potential conflict of interest was reported by the author(s).

References

Agawu, Kofi. 2003. *Representing African Music: Postcolonial Notes, Queries, Positions*. New York: Routledge.

Amiot, Emmanuel. 2007. "David Lewin and Maximally Even Sets." *Journal of Mathematics and Music* 1 (3): 157–172.

Amiot, Emmanuel. 2013. "The Torii of Phases." In *Mathematics and Computation in Music*. Vol. 7937 of *Lecture Notes in Artificial Intelligence*, edited by Jason Yust, Jonathan Wild, and John Ashley Burgoyne, 1–18. Springer.

Amiot, Emmanuel. 2016. *Music Through Fourier Space: Discrete Fourier Transform in Music Theory.* Heidelberg: Cham: Springer.

Amiot, Emmanuel. 2017a. "Interval Content vs. DFT." In *Mathematics and Computation in Music. Lecture Notes in Computer Science*, edited by Octavio A. Agustìn-Aquino, Emilio Luis-Puebla, and Marianna Monteil. Heidelberg: Springer.

Amiot, Emmanuel. 2017b. "Of Decompositions of Nil Sums of Roots of Unity." *Journal of Mathematics and Music* 11 (2): 67–75.

Bach, Carl Philipp Emanuel. 1949. *Essay on the True Art of Playing Keyboard Instruments.* New York: W.W. Norton.

Biamonte, Nicole. 2014. "Formal Functions of Metric Dissonance in Rock Music." *Music Theory Online* 30 (2). http://www.mtosmt.org/issues/mto.14.20.2/mto.14.20.2.biamonte.php.

Bigo, Louis, Moreno Andreatta, Jean-Louis Giavitto, Olivier Michel, and Antoine Spicher. 2013. "Computation and Visualization of Musical Structures in Chord-Based Simplicial Complexes." In *Mathematics and Computation in Music*, Vol. 7937 of *Lecture Notes in Artificial Intellegence*, edited by Jason Yust, Jonathan Wild, and John Ashley Burgoyne, 38–51. Heidelberg: Springer.

Bigo, Louis, Daniel Ghisi, Antoine Spicher, and Moreno Andreatta. 2015. "Representation of Musical Structures and Processes in Simplicial Chord Spaces." *Computer Music Journal* 39 (3): 9–24.

Callender, Clifton, Ian Quinn, and Dmitri Tymoczko. 2008. "Generalized Voice-Leading Spaces." *Science* 320 (5874): 346–348.

Cantazaro, Michael. 2011. "Generalized Tonnetze." *Journal of Mathematics and Music* 5 (2): 117–139.

Clampitt, David, and Thomas Noll. 2011. "Modes, the Height-Width Duality, and Handschin's Tone Character." *Music Theory Online* 17 (1). http://www.mtosmt.org/issues/mto.11.17.1/mto.11.17.1.clampitt_and_noll.html.

Clough, John, and Jack Douthett. 1991. "Maximally Even Sets." *Journal of Music Theory* 35 (1/2): 93–173.

Cohn, Richard. 1997. "Neo-Riemannian Operations, Parsimonious Triads, and Their 'Tonnetz' Representations." *Journal of Music Theory* 41 (1): 1–66.

Cohn, Richard. 1998. "Square Dances with Cubes." *Journal of Music Theory* 42 (2): 283–296.

Cohn, Richard (2012). *Audacious Euphony: Chromaticism and the Triad's Second Nature*, Oxford Studies in Music Theory. Oxford University Press.

Cohn, Richard. 2016. "A Platonic Model of Funky Rhythms." *Music Theory Online* 22 (2). http://mtosmt.org/issues/mto.16.22.2/mto.16.22.2.cohn.html.

Cuddy, Lola L., and Betsy Badertscher. 1987. "Recovery of the Tonal Hierarchy: Some Comparisons Across Age and Levels of Musical Experience." *Perception & Psychophysics* 41 (6): 609–620.

Douthett, Jack. 1997. "3-Dimensional Tiling." Unpublished Letter to John Clough Working Group.

Douthett, Jack, and Peter Steinbach. 1998. "Parsimonious Graphs: A Study in Parsimony, Contextual Transformations, and Modes of Limited Transposition." *Journal of Music Theory* 42 (2): 241–63.

Giavitto, Jean-Louis, and Antoine Spicher. 2016. "A Topological Approach of Musical Relationships." In *Mathemusical Conversations: Mathematics and Computation in Music Performance and Composition*, Vol. 32 of *Lecture Notes Series, Institute for Mathematical Sciences, University of Singapore* edited by Assayag G. Chew, E. and J.B.L. Smith, 283–298. Singapore: World Scientific Press.

Gollin, Edward. 1998. "Some Aspects of Three-Dimensional 'Tonnetze'." *J. Mus. Th.* 42 (2): 195–206.

Gollin, Edward. 2011. "From Matrix to Map: *Tonbestimmung*, the *Tonnetz*, and Riemann's Combinatorial Conception of Interval." In *Oxford Handbook of Neo-Riemannian Music Theories*, edited by E. Gollin and A. Rehding, Oxford Handbooks, 271–293. New York: Oxford University Press.

Gómez-Martín, Francisco, Perouz Taslakian, and Gottfried Toussaint. 2009. "Structural Properties of Euclidean Rhythms." *Journal of Mathematics and Music* 3 (1): 1–14.

Harrison, Daniel. 2002. "Nonconformist Notions of Nineteenth-Century Enharmonicism." *Music Analysis* 21 (2): 115–160.

Hook, Julian. 2011. "Spelled Heptachords." In *Mathematics and Computation in Music*, Vol. 6726 of *Lecture Notes in Computer Science* edited by Carlos Agon, Moreno Andreatta, Gérard Assayag, Emmanuel Amiot, Jean Bresson, and John Mandereau, 84–97. Heidelberg: Springer.

Huron, David. 2006. *Sweet Anticipation: Music and the Psychology of Expectation.* Cambridge, MA: MIT Press.

Krumhansl, Carol. 1990. *Cognitive Foundations of Musical Pitch.* New York: Oxford University Press.

Krumhansl, Carol L., and Lola L. Cuddy. 2010. "A Theory of Tonal Hierarchies in Music." In *Music Perception*, Vol. 36 of *Springer Handbook of Auditory Research*, edited by M. Riess Jones, R. Fay, and A. Popper, 51–87. New York: Springer.

London, Justin. 2002. "Some Non-Isomorphisms Between Pitch and Time." *Journal of Music Theory* 46 (1–2): 127–151.

Mazzola, Guerino. 2002. *The Topos of Music: Geometric Logic of Concepts, Theory, and Performance.* Basel: Birkhäuser Verlag.

Meredith, David. 2011. "Tonal Scales and Minimal Simple Pitch Class Cycles." In *Mathematics and Computation in Music*, Vol. 6726 of *Lecture Notes in Computer Science*, edited by Carlos Agon, Moreno Andreatta, Gérard Assayag, Emmanuel Amiot, Jean Bresson, and John Mandereau, 165–179. Heidelberg: Springer.

Milne, Andrew J., David Bulger, and Steffen A. Herff. 2017. "Exploring the Space of Perfectly Balanced Rhythms and Scales." *Journal of Mathematics and Music* 11 (2): 101–133.

Noll, Thomas. 2018. "One Note Samba: Navigating Notes and their Meanings within Modes and Exo-Modes." In *Mathematical Music Theory: Algebraic, Combinatorial, Topological and Applied Approaches to Understanding Musical Phenomena*, edited by Mariana Montiel and Robert Peck, 109–145. Hackensack, NJ: World Scientific.

Osborn, Brad. 2014. "Kid Algebra: Radiohead's Euclidean and Maximally Even Rhythms." *Perspectives of New Music* 52 (1): 81–105.

Perle, George. 1977. "Berg's Master Array of Interval Cycles." *Musical Quarterly* 63 (1): 1–30.

Quinn, Ian. 2006. "General Equal-Tempered Harmony: Parts Two and Three." *Perspectives of New Music* 45 (1): 4–63.

Rings, Steven. 2011. "Riemannian Analytical Values, Neo- and Paleo-." In *Oxford Handbook of Neo-Riemannian Music Theories*, edited by E. Gollin and A. Rehding, Oxford Handbooks, 478–511. New York: Oxford University Press.

Straus, Joseph N. 2016. *Introduction to Post-Tonal Theory*. 4th ed. New York: W.W. Norton.

Temperley, David. 2007. *Music and Probability*. Cambridge, MA: MIT Press.

Toussaint, Gottfried T. 2013. *The Geometry of Musical Rhythm: What Makes a "Good" Rhythm Good?* Boca Raton, Fl: CRC Press.

Tymoczko, Dmitri. 2005. "Voice-Leading As Generalized Key Signatures." *Music Theory Online* 11 (4). https://mtosmt.org/issues/mto.05.11.4/mto.05.11.4.tymoczko.pdf

Tymoczko, Dmitri. 2011. *Geometry of Music: Harmony and Counterpoint in the Extended Common Practice*. New York: Oxford University Press.

Tymoczko, Dmitri. 2012. "The Generalized Tonnetz." *Journal of Music Theory* 56 (1): 1–52.

Tymoczko, Dmitri, and Jason Yust. 2019. "Fourier Phase and Pitch-Class Sum." In *Mathematics and Computation in Music*, edited by Francisco Martin Gómez, Mariana Montiel, and Octavio A. Agustín-Aquino, Lecture Notes in Computer Science. Heidelberg: Springer.

Yust, Jason. 2015a. "Applications of DFT to the Theory of Twentieth-Century Harmony." In *Mathematics and Computation in Music*, Vol. 9110 of *Lecture Notes in Computer Science*, edited by Tom Collins, David Meredith, and Anja Volk, 207–18. Heidelberg: Springer.

Yust, Jason. 2015b. "Schubert's Harmonic Language and Fourier Phase Space." *Journal of Music Theory* 59 (1): 121–181.

Yust, Jason. 2016. "Special Collections: Renewing Set Theory." *Journal of Music Theory* 60 (2): 213–262.

Yust, Jason. 2017a. "Harmonic Qualities in Debussy's 'Les sons et les parfums tournent dans l'air du soir.'" *Journal of Mathematics and Music* 11 (2–3): 155–173.

Yust, Jason. 2017b. "Probing Questions about Keys." In *Mathematics and Computation in Music*, edited by Octavio A. Agustìn-Aquino, Emilio Luis-Puebla, and Marianna Monteil, Lecture Notes in Computer Science. Heidelberg: Springer.

Yust, Jason. 2018a. "Geometric Generalizations of the Tonnetz and their Relation to Fourier Phase Spaces." In *Mathematical Music Theory: Algebraic, Combinatorial, Topological and Applied Approaches to Understanding Musical Phenomena*, edited by Mariana Montiel and Robert Peck, 253–278. Boca Raton, Fl: World Scientific.

Yust, Jason. 2018b. *Organized Time: Rhythm, Tonality, and Form*. New York: Oxford University Press.

Yust, Jason. 2019. "Stylistic Information in Pitch-Class Distributions." *Journal of New Music Research* 48 (3): 217–231.

Yust, Jason. 2002. "Ganymede's Heavenly Descent." *Music Analysis* 39 (1).

Žabka, Marek. 2014. "Dances with the Scales: Subchromatic Tone Systems." *Journal of Music Theory* 58 (2): 179–233.

Homological Persistence in Time Series: An Application to Music Classification

Mattia G. Bergomi and Adriano Baratè

Meaningful low-dimensional representations of dynamical processes are essential to better understand the mechanisms underlying complex systems, from music composition to learning in both biological and artificial intelligence. We suggest to describe time-varying systems by considering the evolution of their geometrical and topological properties in time, by using a method based on persistent homology. In the static case, persistent homology allows one to provide a representation of a manifold paired with a continuous function as a collection of multisets of points and lines called persistence diagrams. The idea is to fingerprint the change of a variable-geometry space as a time series of persistence diagrams, and afterwards compare such time series by using dynamic time warping. As an application, we express some music features and their time dependency by updating the values of a function defined on a polyhedral surface, called the *Tonnetz*. Thereafter, we use this time-based representation to automatically classify three collections of compositions according to their style, and discuss the optimal time-granularity for the analysis of different musical genres.

1. Introduction

A time-varying system can be interpreted as a series of relevant geometric and topological events. Persistent homology has mainly been applied to the study of static point clouds and shapes (Edelsbrunner and Harer 2009), by providing a description of both the geometry and topology of the analyzed space. One of the reasons that make persistent homology so effective when applied to the study of static spaces is that it provides a representation in which the features of the space appear arranged by relevance. Thus, the analysis can be tuned on a specific application need, by balancing the computational cost and the retrieval of details. Briefly one can think about persistent homology as a scalable fingerprint of a static object.

Persistent homology has been generalized to time-varying systems either by considering continuous representations (Cohen-Steiner, Edelsbrunner, and Morozov 2006), or introducing statistics, in order to evaluate the evolution in time of the analyzed system (Munch 2013; Turner et al. 2014).

A time series is a collection of values obtained through subsequent observations in time. Time-series data mining (Esling and Agon 2012) is an attempt to organize data, in furtherance of visualizing the *contour* of data, avoiding negligible details and creating a consistent, interpretable representation. Due to their generality and flexibility, time series are extensively used in applications, e.g. classification (Bakshi and Stephanopoulos 1994), segmentation (Keogh et al. 2004), and supported by a strong theoretical framework (Keogh and Kasetty 2003).

We suggest to combine the scalability (among other properties) of persistent homology and the notion of time series, in order to characterize the evolution of a variable-geometry space in time. We will discuss how this representation allows one to distinguish between relevant and noisy observations by utilizing the natural notion of distance defined in the context of persistent homology as a cost function (Senin 2008). In addition, the computation of the dissimilarity between time series (Liao 2005) will allow us to compare several time-varying systems; i.e. to find the timespans (if they exist) where two time series can be considered comparable in agreement with their time-dependent geometric and topological characterization.

As an application, we consider a variable-geometry polyhedral surface T inspired by the *Tonnetz*, a graph widely used in computational musicology to represent the relationships among notes. We will perform automatic stylistic clustering of three collections of classical, jazz, and pop songs, by updating the geometry of T according to the information deduced from a certain composition, and therefore computing the associated time series of persistence diagrams.

2. Persistent homology background

In computational topology (Edelsbrunner and Harer 2009), persistent homology provides a multiscale description of the topology of a given space. The efficacy and versatility of this technique is demonstrated by its numerous and diverse applications. See Ferri (2017) for a review. In the remainder of this section, we will provide a basic intuition of homology and persistent homology.

2.1. *Homology*

Homology is a standard theory in Algebraic Topology first described in Poincaré (1895). Here we will give an intuition about homology and limit ourselves to discuss the basic concepts and definitions needed in the remainder of this work. However, homology can be defined in more general settings. We refer the reader to Munkres (1984) and Hatcher (2002) for a detailed, rigorous description.

The extremely simplified view on homology needed for the understanding of what follows is that it allows one to map a topological space X (think of a manifold for simplicity) to a sequence of algebraic objects, here vector spaces denoted by $H_i(X)$ for every $i \in \mathbb{Z}$. Although $H_i(X)$ are not always vector spaces, it is possible to achieve this structure by computing homology with coefficients in a field. We compute homology in $\mathbb{Z}/2\mathbb{Z}$. See Munkres (1984), Hatcher (2002) and Bergomi (2015) (Part 1, Chapter 2), for more details. The dimension of the vector space $H_i(X)$ counts the number of i-dimensional *holes* of X. Thus, $\dim(H_0(X))$ corresponds to the number of connected components of X, the dimension of $H_1(X)$ to the number of holes, $H_2(X)$ voids, and so on. We define the i-Betti number associated with X as $\beta_i(X) = \dim(H_i(X))$, or simply β_i when no confusion arises. See Figure 1 for an example.

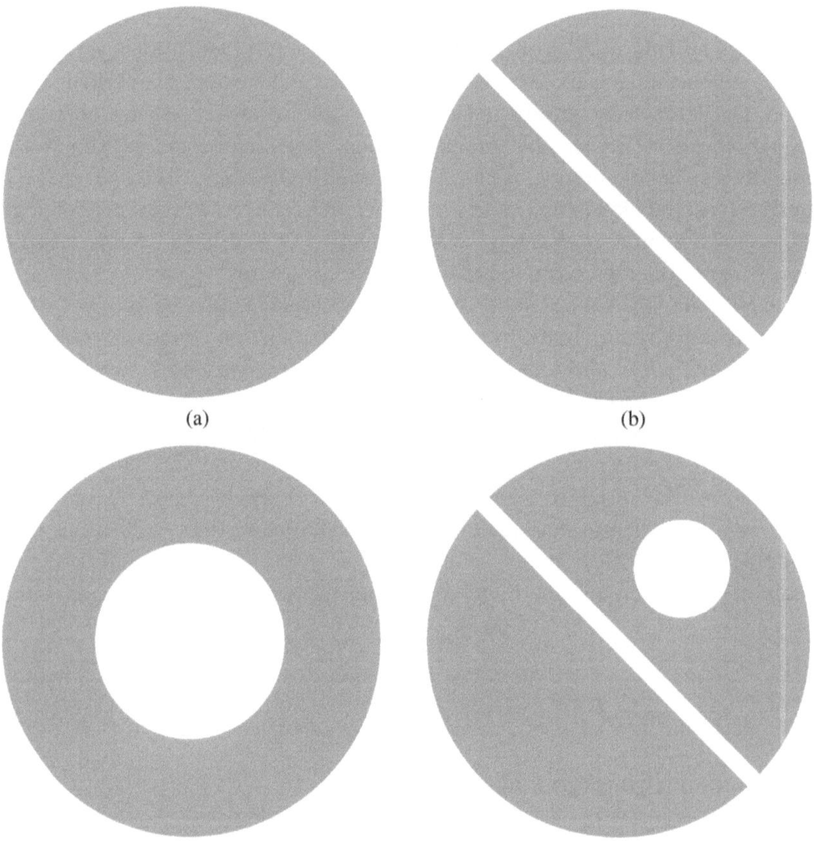

(a) (b) (c) (d)

Figure 1. An intuition on Betti numbers. For each of the four example spaces, the 0-Betti number β_0 counts the number of connected components, while β_1 counts 1-dimensional holes. (a) $\beta_0 = 1, \beta_1 = 0$. (b) $\beta_0 = 2, \beta_1 = 0$. (c) $\beta_0 = 1, \beta_1 = 1$ and (d) $\beta_0 = 2, \beta_1 = 1$.

2.2. *Persistent homology*

Clearly, the computation of the Betti numbers associated with a certain space X causes a huge loss of information. Persistent homology allows us not only to alleviate, but control this phenomenon, by considering the evolution of the Betti numbers on a *filtration* of X, i.e. a sequence of nested spaces starting with the empty set and ending with the entire space X.

2.2.1. *Filtering functions*

Let X be a triangulable manifold and $f : X \to \mathbb{R}$ a continuous function. A filtration of X is defined as a collection of nested sublevel sets $X_u = f^{-1}((-\infty, u])$, $u \in \mathbb{R}$. The function f is called a *filtering function* or simply a *filter*.

The choice of f depends on the geometric property of X one wants to take into account. To understand how the choice of a filtering function alters the representation of a given space, it is necessary to introduce the following definition. A real number r is a *regular homological value* for the pair (X,f) if it exists $\varepsilon > 0$ such that for every pair of real numbers $x < y$ in $(r - \varepsilon, r + \varepsilon)$, the homology of X_x is isomorphic to the one of X_y. For the purposes of this manuscript, one can simply think that X_x and X_y have the same Betti numbers. Otherwise r is said to be a homological critical value of (X,f). See Govc (2013) for details. A filtering function with a finite number of homological critical values $\{a_1,\ldots,a_n\}$, finite i-Betti numbers for each sublevel set X_{a_k}, $k \in \{1,\ldots,n\}$ and $i \in \mathbb{Z}$ is said to be *tame*.

We can now consider the filtration

$$\emptyset \subseteq X_{a_1} \subseteq X_{a_2} \subseteq \cdots \subseteq X_{a_n} = X,$$

compute the Betti numbers associated with each sublevel set, and observe their evolution throughout the filtration. The scale at which a feature is significant is measured by considering its longevity (i.e. persistence) along the filtration.

2.2.2. *Persistence diagrams*

All the information carried by this analysis can be summarised in a simple and computationally efficient form. We do this by enumerating the values of birth and death of each component that produces a change in the Betti numbers along the filtration. Let us denote each of these points by (b_i, d_i). The multiset (a set whose elements can be repeated) $D = \{(b_i, d_i)\}_{i\in\{1,\ldots,m\}}$ is called *persistence diagram*. Graphically, a persistence diagram is represented as a collection of points in $\mathbb{R}^2$ with the following characteristics:

(1) $d_i \geq b_i$ for every $(b_i, d_i) \in D$. This condition means that a n-dimensional hole cannot die before it is born.
(2) The ordinate d_i can be infinite. It is indeed possible that a n-dimensional hole that is born along the filtration never dies.

The points of the persistence diagram are called *cornerpoints*; points of the form (b_i, ∞) with infinite lifespan are said to be *cornerlines* and graphically rendered as vertical half lines. The distance of a cornerpoint from the diagonal is its *persistence*. See Figure 2(a,b) for an example. In the figure, observe how the local minima a_2 and a_4, when merging with the maxima a_3 and a_5 give rise to cornerpoints with different persistence, because $a_3 - a_2 < a_5 - a_4$. Finally, observe that more cornerpoints can have the same coordinates. This generally reflects the presence of symmetries of the analyzed space with respect to the filtering function. Set of overlapping cornerpoints are considered as a single cornerpoint with multiplicity equal to the cardinality of the set.

2.2.3. *Bottleneck distance*

The *bottleneck distance* (d'Amico, Frosini, and Landi 2006) is naturally a metric between persistence diagrams defined as

$$d_B(D,D') = \min_{\sigma} \max_{p\in D} d(p,\sigma(p)),$$

where D and D' are two kth persistence diagrams, σ varies among all the bijections between D and D', and

$$d\left((u,v),(u',v')\right) = \min\left\{\max\left\{|u-u'|,|v-v'|\right\},\max\left\{\frac{v-u}{2},\frac{v'-u'}{2}\right\}\right\},$$

for every $(u,v),(u',v') \in \{(x,y) \in \mathbb{R}^2 : x \leq y\}$.

The bottleneck distance is the most used metric to compare persistence diagrams, because of its properties of optimality (d'Amico, Frosini, and Landi 2010) and stability (Cohen-Steiner,

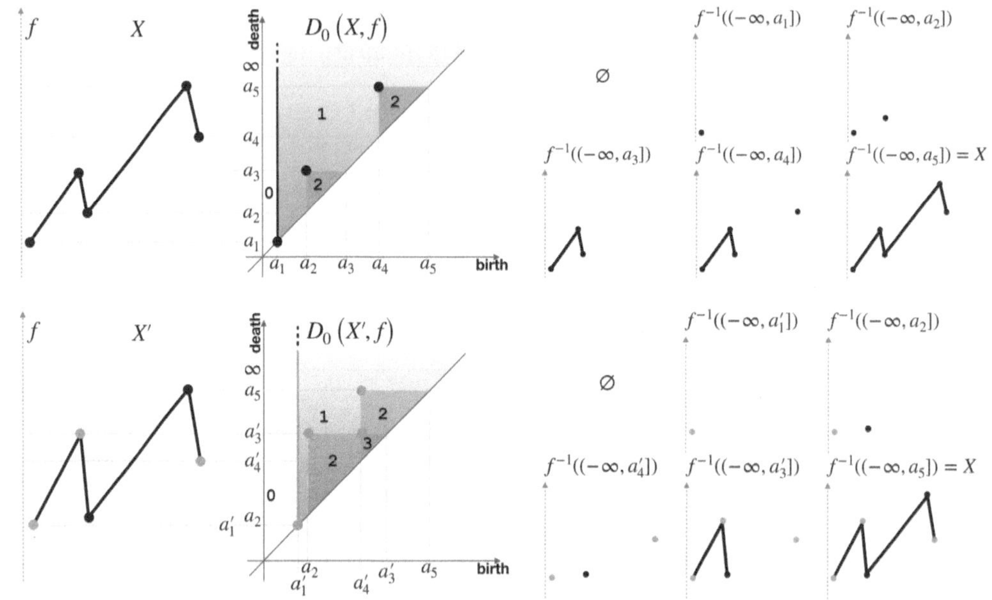

(a) (b)

Figure 2. Panel (a): persistent homology induced by the sublevel sets of the height function f on the space X. $\{a_i\}_{i\in\{1,\ldots,5\}}$ are the ordinates of local maxima and minima of $f : X \to \mathbb{R}$. The 0-persistence diagram on the right of the panel shows the evolution of the 0-Betti number (number of connected components) throughout the filtration $f^{-1}((-\infty, a_i])$. The integer numbers reported in the persistence diagram correspond to the value of the 0-persistent Betti number function for every $(u, v) \in \mathbb{R}^2$. For instance, before the sublevel set $f^{-1}((-\infty, a_1])$ is reached, no connected component exists. For values $a_2 \leq f(x) < a_3$ two connected components exist. Those will merge in a single connected component once a_3 is reached. The cornerline of coordinates (a_1, ∞) is associated with the connected component corresponding to the absolute minimum of f with value a_1. Each step of the filtration is depicted in Panel (b). Panel (c): a variation X' (in orange) of the space X of Panel (a). The variation is detected by the 0-persistence diagram and highlighted in orange. The filtration associated with (X', f) is illustrated in Panel (d). (a) 0-persistence diagram. (b) Sublevel set filtration of X with respect to f. (c) A variation X' of the space X and its 0-persistence diagram and (d) Sublevel set filtration of X' with respect to f.

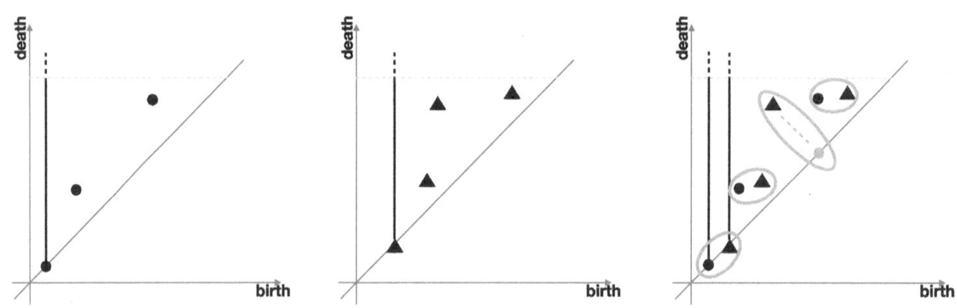

(a) (b)

Figure 3. (a) Two persistence diagrams. (b) Cornerpoints are matched by choosing to the bijection that minimises the sum of the distances. Observe how one of the cornerpoints represented as triangles is matched with its projection on the diagonal. (a) Persistence diagrams and (b) Matching.

Edelsbrunner, and Harer 2007). See Figure 3 for an intuition. It should be clear from the figure that in order to measure the distance between two persistence diagrams with a different number of cornerpoints, it is necessary to consider bijections that map some cornerpoints placed on the diagonal, i.e. with zero persistence.

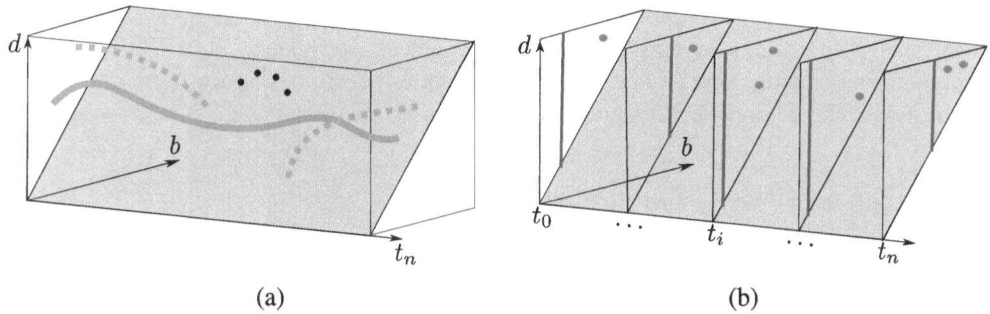

Figure 4. (Left) An example of vineyard. The axes represent the time t, the birth level b and the death level d of the k-homology classes of a system evolving in time. Vines are represented as continuous paths. (Right) Persistence diagrams associated with evenly spaced observations of a time-varying system.

2.3. *Time-varying systems*

A change of the geometry of a shape corresponds to an update of its persistence diagrams. This update encodes the relevance of change in time, with respect to the filtering function. See Figure 2.

Persistent homology has been generalized to the study of time-varying systems in Cohen-Steiner, Edelsbrunner, and Morozov (2006). Let $f, g : \mathbb{X} \to \mathbb{R}$ be tame functions, $I = [0, 1] \subset \mathbb{R}$ and $H : \mathbb{X} \times I \to \mathbb{R}$ a homotopy between f and g, i.e. a continuous deformation between f and g. See Hatcher (2002) for more details on homotopy theory. Assuming that $H(x, t)$ is tame for every $t \in I$, we obtain a k-persistence diagram for every pair $(t, k) \in I \times \mathbb{Z}$, named a *vineyard*. In this representation, each cornerpoint is associated with a trajectory called a *vine* (cornerlines can be seen as cornerpoints (u, v), such that $v \gg u$). See Figure 4 for an example.

3. Persistence time series

Vineyards are powerful tools for describing time-varying systems. Indeed, given a time-varying system (X, t) they encode the changes in time of the homological critical values induced by a given filtering function. However, their interpretation is not intuitive, and there is not a general technique that allows one to compare vineyards associated with two time-varying systems.

Let $\mathbb{X}$ be a topological space, $f : \mathbb{X} \times I \to \mathbb{R}$ a tame function for every $t \in I$, and $\mathbf{t} = \{ t_0, \ldots, t_n \}$ a set of $(n + 1)$ evenly spaced points of I. A k-persistence diagram $D_k(X, f)_i$ is associated with every $t_i \in \mathbf{t}$. The collection of these *k-persistence snapshots* is a time series $\mathcal{D}_{k,n}(X, f) = \{ D_k(X, f)_i \}_{i=0}^{n} \subset D_\infty$, where (D_∞, d_B) is the space of persistence diagrams equipped with the bottleneck distance. We name $\mathcal{D}_{k,n}(X, f)$, or simply $\mathcal{D}(X, f)$, a *k-persistence time series*.

There exist several methods to evaluate the dissimilarity of two time series (Liao 2005; Esling and Agon 2012; Lines and Bagnall 2015). Here, we will consider Dynamic Time Warping (DTW) (Berndt and Clifford 1994; Senin 2008). One of the most important features of DTW is that it allows us to align, and hence measure the dissimilarity between time series of different lengths, by providing their global optimal warping.

3.1. *Dynamic time warping*

Let $X = \{x_1, \ldots, x_n\}$ and $Y = \{y_1, \ldots, y_m\}$ be two time series, where $n, m \in \mathbb{N}$. The first ingredient needed for the computation of the optimal warping path between X and Y is a measure of dissimilarity between pairs of observations. The DTW algorithm will align the two time series by rearranging their points, in order to minimise their global dissimilarity.

Let us assume that the observations of both time series belong to a space Φ, said to be the *feature space*. The dissimilarity function, or *cost function*, is defined as $c : \Phi \times \Phi \to \mathbb{R}$, such that

(1) $c(x, y) \geq 0$ for all observations $x, y \in \Phi$;
(2) $c(x, y) = 0$ if and only if $x = y$.

Observe that a cost function is not necessarily a metric.

The second step is the construction of the *local cost matrix* $C \in \mathbb{R}^{n \times m}$, whose entries are defined as $C_{ij} := c(x_i, y_j)$, for $i \in \{1, \ldots, n\}$ and $j \in \{1, \ldots, m\}$. A warping path $\gamma = (\gamma_1, \ldots, \gamma_l)$ is a sequence of pairs of indices $\gamma_k = (i_k, j_k)$, where $i_k \in \{1, \ldots, n\}$, $j_k \in \{1, \ldots, m\}$ and $k \in \{1, \ldots, l\}$, such that the following conditions hold:

(1) $\gamma_1 = (1, 1)$ and $\gamma_l = (m, n)$ for every $l \in \mathbb{N}$. This condition simply states that the starting and ending observations of the two time series have to be aligned, i.e. we are aligning the two time series globally.
(2) Given $\gamma_k = (i_k, j_k)$ and $\gamma_{k+1} = (i_{k+1}, j_{k+1})$, we have $i_k \leq i_{k+1}$ and $j_k \leq j_{k+1}$. This condition ensures that the natural ordering induced by time on the observations is preserved.
(3) The difference between two subsequent steps in the warping path $\gamma_{k+1} - \gamma_k \in \{(0, 0), (1, 0), (0, 1), (1, 1)\}$. This last condition constrains the size of the shifts in time used to align the two time series.

The *total cost* of a (n, m)-warping path γ over the observations of X and Y is defined as

$$c_\gamma(X, Y) := \sum_{k=1}^{l} c(x_{i_k}, y_{j_k}).$$

An *optimal warping path* on X and Y is a warping path realising the minimum total cost. We are now ready to define the *DTW distance* between X and Y:

$$DTW(X, Y) := \min \left\{ c_\gamma(X, Y) \;\middle|\; \gamma \text{ is a} \right\} (n, m)\text{-warping path}.$$

Note that the minimum always exists because the set is finite. In Figure 5, a simple example of global alignment of time series *via* DTW is depicted.

3.2. *DTW for persistence time series*

When considering persistence time series, the bottleneck distance can be used as a cost function. Let $\mathcal{D}_{k,n}(X, f)$ and $\mathcal{D}_{k,m}(Y, g)$ be two kth persistence time series. By definition, the bottleneck distance satisfies the properties that characterize a cost function. The DTW between k-persistence time series is given by an optimal warping path defined as:

$$\begin{aligned}&\mathrm{DTW}\,(\mathcal{D}(X,f), \mathcal{D}(Y, g))\\ &\quad = \min \left\{ d_{B_\gamma}\,(\mathcal{D}(X,f), \mathcal{D}(Y, g)) \;\middle|\; \gamma \text{ is an } (n, m) - \text{ warping path} \right\}.\end{aligned}$$

The DTW inherits the symmetry by the bottleneck distance, and the tameness of f and g assures stability of every diagram $D(X, f_i)$ and $D(Y, g_j)$, for every $i \in \{1, \ldots, n\}$ and $j \in \{1, \ldots, m\}$.

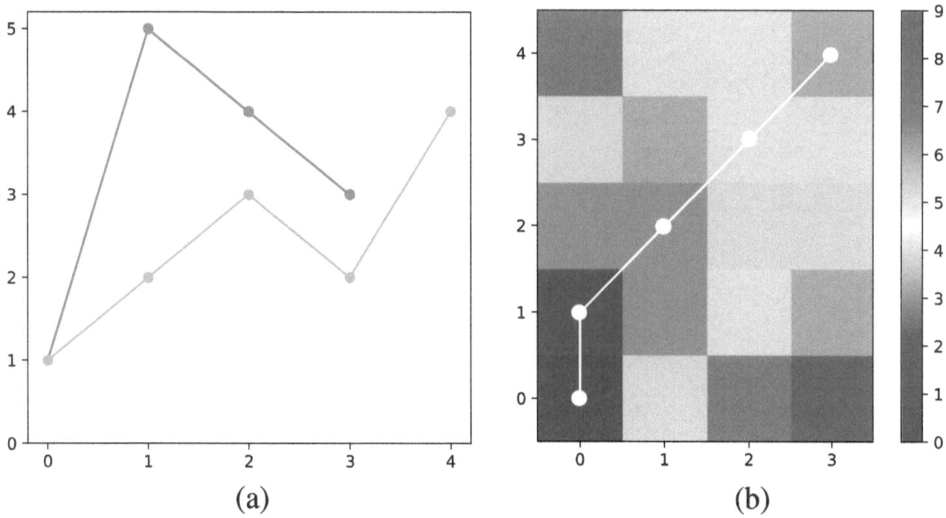

Figure 5. (a) Two time series of length 5 and 4, respectively. (b) The cost matrix and optimal warping path between the two time series.

Consider the prefix sequences $\mathcal{D}_{k,\alpha}(X,f)$ and $\mathcal{D}_{k,\beta}(Y,g)$, where $\alpha \leq n$ and $\beta \leq m$ are natural numbers. The *accumulated cost matrix* is defined as

$$A(\alpha,\beta) := DTW\left(\mathcal{D}_{k,\alpha}(X,f), \mathcal{D}_{k,\beta}(Y,g)\right).$$

Then $A(n,m) = DTW\left(\mathcal{D}_{k,n}(X,f), \mathcal{D}_{k,m}(Y,g)\right)$.

4. Musical setting

Music, or at least some of its features, can be represented using a variable-geometry space. Before illustrating our model, it is necessary to introduce some basic definitions.

A *note in equal tuning*, denoted by $\mathfrak{n}$, can be modeled as a pair $(p,d) \in \mathbb{R}^2$, where p is its *pitch*, and d its *duration* in seconds. In particular, if ν denotes the fundamental frequency of $\mathfrak{n}$, we have $p(\nu) = 69 + 12\log_2(\nu/440)$, where 440 Hz is the fundamental frequency of the A of the fourth octave of the piano (Cheveignéde 2005).

Perceptively, two notes an octave apart are really similar (Burns and Ward 1999), thus, it is a common practice in computational musicology to identify pitches modulo octave, by considering *pitch-classes* $[p] = \{\, p + 12k : k \in \mathbb{Z} \,\} \cong \mathbb{R}/12\mathbb{Z}$. See Figure 6 for a representation of both the pitch and pitch-class spaces.

4.1. *A 3-dimensional, variable-geometry Tonnetz*

Originally, the *Tonnetz* was described as a graph representing the harmonic relationships among pitches, see for instance in Euler (1774). Later, it has been largely redefined, and generalized to several formalisms (Cohn 1997; Žabka 2009; Douthett and Steinbach 1998). We will consider the simplicial complex interpretation of the *Tonnetz* originally given in Bigo et al. (2013). In this setting, the *Tonnetz* is modeled as an infinite planar simplicial complex, whose 0-simplices (i.e. vertices) are labeled with pitch-classes, 1-simplices (i.e. edges) connect vertices whose labels form either perfect fifth, major, and minor third intervals. The labels associated with the vertices

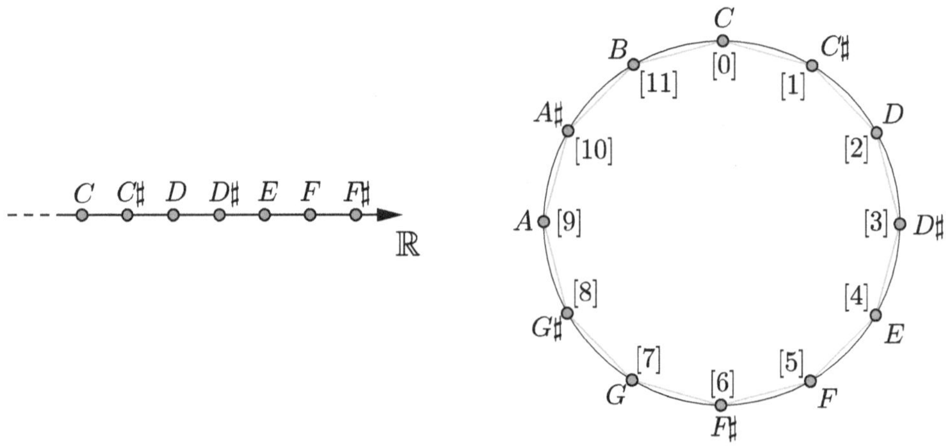

Figure 6. Fundamental music representation spaces. On the left pitches are visualized as points on the real line. The pitch-class space is visualized on the right.

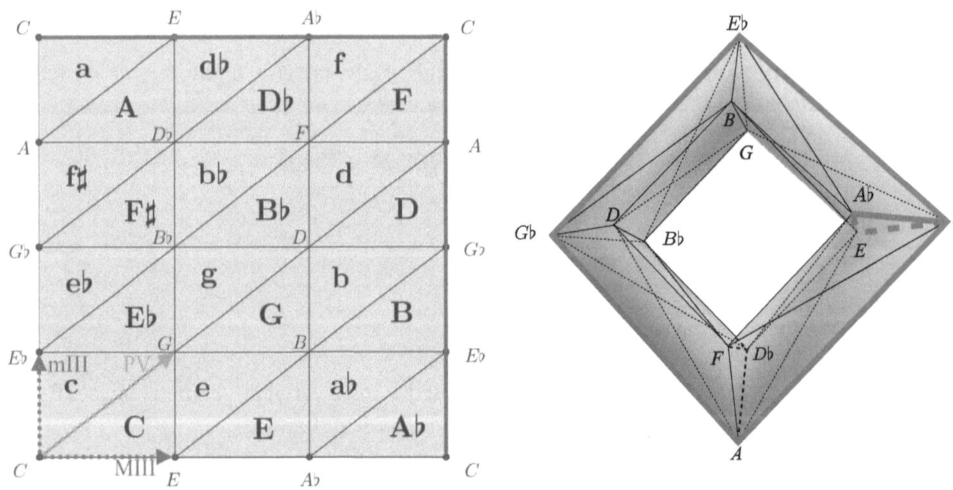

Figure 7. On the left, a finite subcomplex of the planar *Tonnetz* T. On the right, the *Tonnetz* torus $\mathbb{T}$.

of 2-simplices (i.e. triangles) correspond to either major or minor triads. A finite subcomplex of the *Tonnetz* T is depicted in Figure 7(left). We observe that the labels on its vertices are periodic with respect to the transposition of both minor and major third. This allows one to generate the finite toroidal representation $\mathbb{T}$ displayed in Figure 7(right).

It is possible to analyze and classify music by considering the subcomplexes of T generated by a sequence of pitch-classes (Bigo et al. 2013). However, this approach does not allow us to discriminate musical styles in a geometric or topological sense. Indeed, as the example in Figure 8 shows, two perceptively distinct sonorities can be represented by isomorphic subcomplexes.

Imagine the planar Tonnetz as embedded in $\mathbb{R}^3$. Then, the information concerning the harmonic relationships and the *temporal* hierarchy (durations) of notes in a musical phrase can be expressed by displacing the vertices of the *Tonnetz* along the axis normal to its surface. Moreover, music is often organised in bars. For instance, modulations generally occur every four or eight bars in a jazz context, and oftentimes melodic lines are arranged in a question and answer

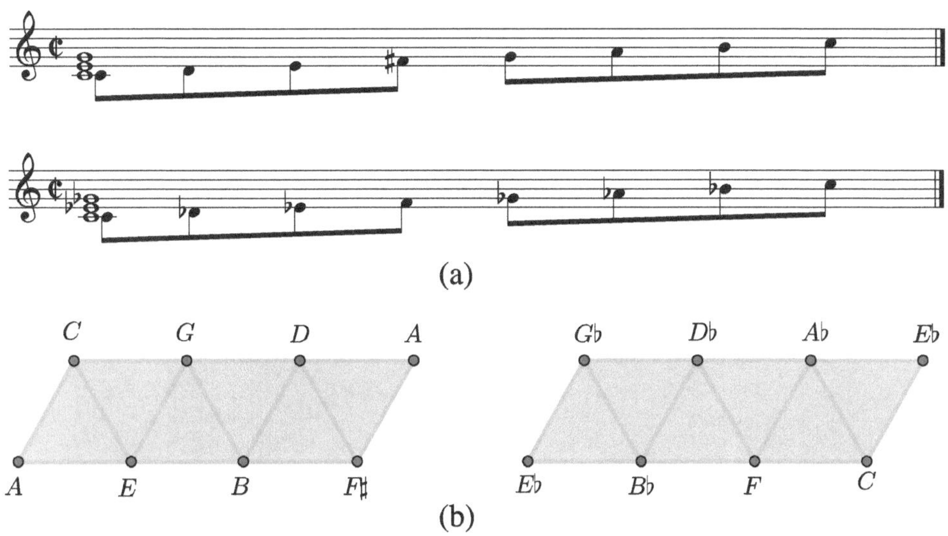

Figure 8. Two different modes represented by isomorphic subcomplexes of the *Tonnetz*. The subcomplexes are built by considering the simplices whose vertices and edges form a 2-clique community (triangles sharing an edge), such that vertices are labelled with all the pitches belonging to the chosen modal scale. (a) The lydian mode. (b) The locrian mode. (c) Ionian subcomplex and (d) Locrian subcomplex.

paradigm consisting of cycles of two or four bars in pop music. The idea is to take into account this natural segmentation to create a windowing of a given composition.

In the following paragraphs, we will describe how to create deformed versions of the planar Tonnetz by considering the pitch-classes and duration of the notes used in a composition. Thereafter, we will show how the evolution in time of a composition can be taken into account, generating a time-varying Tonnetz.

4.1.1. *Deformation of the Tonnetz*

We start by recalling that in our notation the planar Tonnetz is denoted by T, while $\mathbb{T}$ is the symbol used for its toroidal representation. Let V be the set of vertices of T. Consider the simplest case of a finite collection of notes $P = \{\mathfrak{n}_1, \ldots, \mathfrak{n}_m\} = \{(p_1, d_1), \ldots, (p_m, d_m)\}$, where p and d are the pitch and duration of each note. Assume that $\{\mathfrak{n}_{i_1}, \ldots, \mathfrak{n}_{i_k}\}$, $k \in \mathbb{N}, k \leq m$, is the subset of notes in P whose pitch belongs to the same pitch-class. Then, for every pitch-class $[p]$, we define a function

$$h : V_{[p]} \subset \mathbb{R}^3 \rightarrow \mathbb{R}^3$$

$$(x_v, y_v, 0) \mapsto (x_v, y_v, d_{v_i}),$$

where $V_{[p]} \subset V$ is the subset of vertices of T labeled with $[p]$, $d_{v_i} = \sum_{j=1}^{m} d^i_{l_j}$. The function h sums the durations of notes whose pitch belong to the same pitch-class. This can be visualised as a deformation $\overline{T}$ of the planar Tonnetz as depicted in Figure 9, panel (a). More importantly, by linear extension, it defines an ordering on the simplices of T, that allows us to induce a filtration on the finite simplicial complex $\mathbb{T}$. The filtration on $\mathbb{T}$ is consequently finite, i.e. with a finite number of homological critical values. Always in Figure 9 the configurations of a maximum, a minimum and a saddle on the geometrical realization of a portion of the deformed Tonnetz are depicted. Observe that, in our case a maximum or minimum can be a whole subcomplex of connected pitch-classes, whose vertices share the same height. A saddle can be created by playing

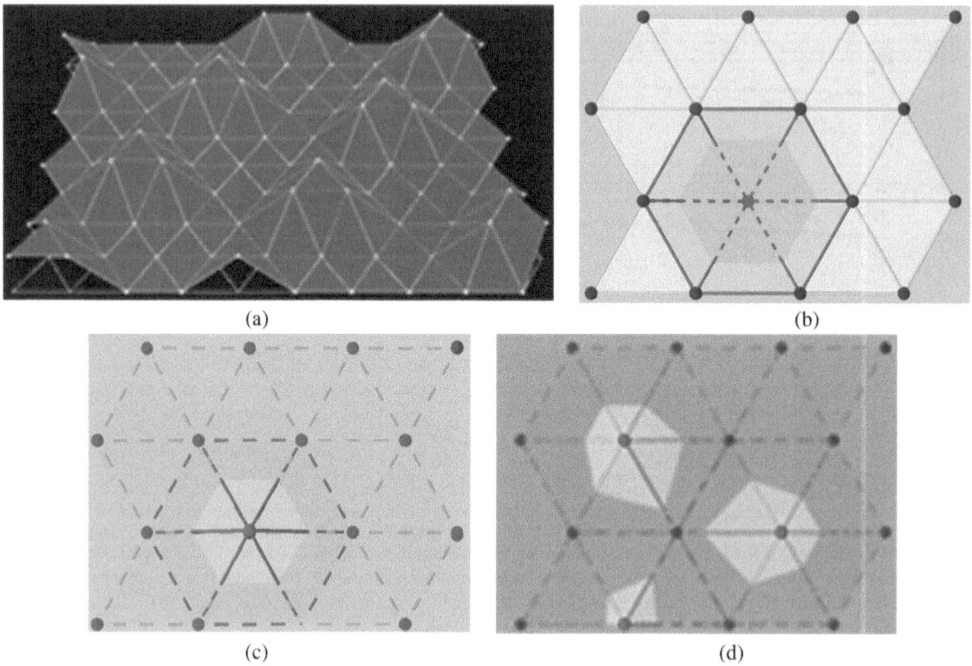

Figure 9. (a) The *Tonnetz* deformed with a major triad that appears as a 2-simplex corresponding to a maximum of the height function. Panels (b–d) represent critical points of the height function defined the deformed Tonnetz and used to induce a filtration on its toroidal counterpart $\mathbb{T}$. (b) Minimum. (c) Maximum and (d) Saddle.

a chromatic cluster of three notes (Cohn 1997). The musical interpretation of these configurations and how they shape the persistence diagrams computed on the filtration induced on $\mathbb{T}$ will be discussed in the following section.

4.1.2. *The time-varying Tonnetz*

Let $Q = \{Q_0, \ldots, Q_s\}$ be a composition consisting of $s + 1$ bars and $w \in \mathbb{N}$ be the number of bars per window we want to consider. Let us denote each window by $P_i = Q_i \cup Q_{i+1} \cup \cdots \cup Q_{i+w-1}$. Then, $Q = \cup P_i$. We can now apply the procedure described in the previous paragraph to every P_i, obtaining a time series whose observations are a deformed configuration of the *Tonnetz*. For the sake of intuition, a 3-dimensional interactive animation showing how the *Tonnetz* is deformed by a musical phrase is available at this link.[1]

The collection of deformed *Tonnetze* $\{\overline{T}_i\}$ associated with a composition will be the core of the following topological analysis. The assumption is that, given a composition, a meaningful and sufficiently refined windowing allows one to generate an accurate fingerprint of its compositional style.

5. Musical interpretation and applications

The geometry and topology of a deformed *Tonnetz* $\overline{T}$ can be considered as musical descriptors (Casey et al. 2008), and analyzed with persistent homology by defining the height function f

[1] https://mgbergomi.github.io/html/def_ton_and_sounds/examples/deformed_tonnetz_int_sound_pers.html

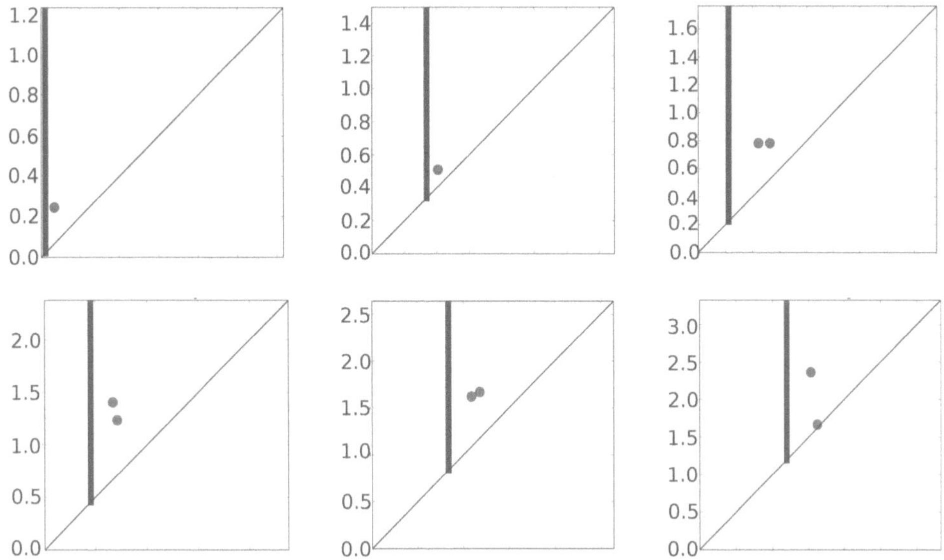

Figure 10. The first six observations of the 0-persistence time series. *Klavierstück I* – Schönberg. Persistence snapshots are taken every 8 bars.

on $\overline{T}$ to induce a lower level set filtration on the torus $\mathbb{T}$. The usage of the height function guarantees the invariance of the persistence diagrams modulo musical transposition, see Bergomi, Baratè, and Di Fabio (2016), where persistent homology has been used to fingerprint compositions as static shapes. However, this static representation does not take into account the time-dependent nature of music. We believe that persistence time series are a more natural and finer tool for the analysis of music. Indeed, this approach makes it possible to detect local relevant phenomena encoded in the progressive geometric update of the deformed *Tonnetz*. In the following applications, we will consider time series whose ith observation consists of the 0-persistence diagram associated with the space $\overline{T}_i$ defined in the previous section.

First of all, it is necessary to provide an interpretation of the music features represented as a persistence diagram.

In Figure 10, a sequence of six 0-persistence diagrams computed considering an 8-bar windowing of Schönberg's *Klavierstück I* is depicted. Consider the first diagram of the first row of the figure: a cornerline reveals the connected nature of $\mathbb{T}$ and represents the absolute minimum of the height function. This minimum corresponds to the subcomplex of the *Tonnetz* that is less used in the composition. The cornerpoint highlights the presence of a second minimum of the height function associated with a subcomplex of $\mathbb{T}$ which is disconnected from the first one. In musical terms, the presence of these two connected components can be associated with the atonal nature of the piece. The lifespan of the cornerpoint measures the relevance of this stylistic feature. The remainder of the observations describe the changes in terms of death and birth levels of these connected components. Moreover, the growth of the birth levels of the points of the whole multiset grabs the homogeneous gain of *height* of the entire simplicial complex, in time. This means that the whole chromatic scale is uniformly used in the composition, both in terms of pitches and duration of the notes. The low relative distance between cornerpoints represents the evenness, in terms of usage, of dissonant intervals during the composition: we recall how a saddle can be formed by playing three notes at a half-tone distance.

We are now ready to align two k-persistence time series, compute their dissimilarity and their optimal warping path.

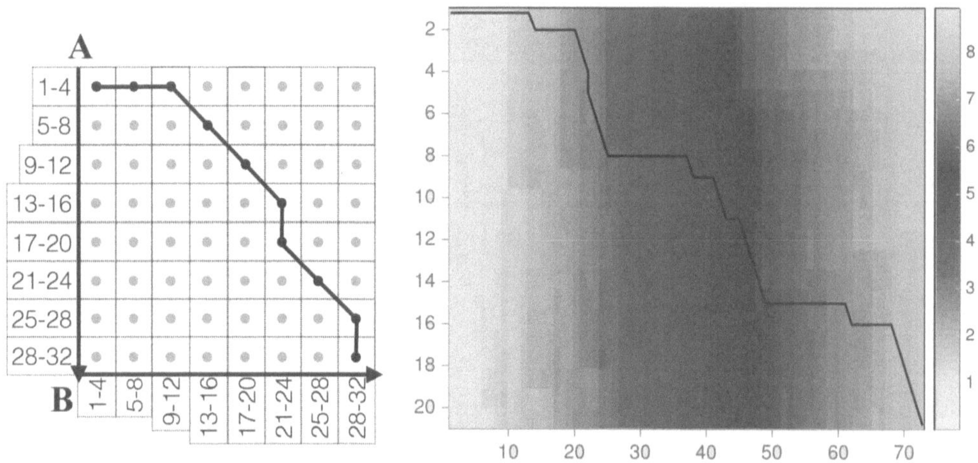

(a) (b)

Figure 11. (Left) Dynamic time warping between persistence time series associated with two compositions A and B. Observations are labelled according to a 4-bars windowing. (Right) Optimal warping path between two versions of *Caravan*. The positions of the gaps correspond to the solo parts of the longer version (frames 25–50 and 51–65, respectively). (a) Optimal warping path and (b) Two versions of Caravan.

5.1. *Optimal warping path*

The first step in the calculation of an optimal warping path, given two kth persistence time series, is the computation of the pairwise bottleneck distance between their observations. The low dimensionality and simple structure of $\mathbb{T}$ assure that this normally computationally hard task can be performed in a reasonable amount of time.

In musical terms, an optimal warping path returns the comparable regions of the two compositions, represented by similar persistence diagrams with respect to the bottleneck distance. In other words, time regions of the compositions that share a similar use of the entire set of pitch-classes (cornerline), or disconnected intervals on the Tonnetz both in terms of relevance (distance from the diagonal of the cornerpoints), and in a balanced or unbalanced way (relative distance and multiplicity of the cornerpoints), are aligned in an optimal warping path. In Figure 11(a), two persistence time series associated with the compositions A and B are represented by piecewise line segments and their observations are labelled according to a 4-bars windowing. The thick line represent an optimal warping path. The first twelve bars of A are associated with the first four bars of B, in the figure an optimal warping path connects the entries $(1, 1)$ and $(1, 3)$ of the accumulated cost matrix. Symmetrically, the two last observations of A are associated with the last one of B. In this region, the warping path is represented by a vertical line segment. Diagonal line segments highlight regions of the two compositions that can be *aligned* with a reasonable cost with respect to the bottleneck distance, and hence, share similar topological features.

In the following applications, we use DTW to compute the dissimilarity between 0-persistence time series associated with three datasets composed by classical, pop, and jazz compositions, respectively. In Figure 12, the dissimilarity scores computed by aligning the compositions belonging to the datasets are depicted.

Classical dataset: Observe the first row of the figure. The pieces of the dataset are listed in Table 1. Proceed by reading the matrix from top to bottom. Both Schönberg's compositions (we will denote them by DK11-1 and DK11-2) gave high dissimilarity score when aligned with the tonal pieces. The first row of the matrix represents the dissimilarity scores computed by comparing DK11-2 with the other compositions of the dataset. The two minimal scores we retrieved

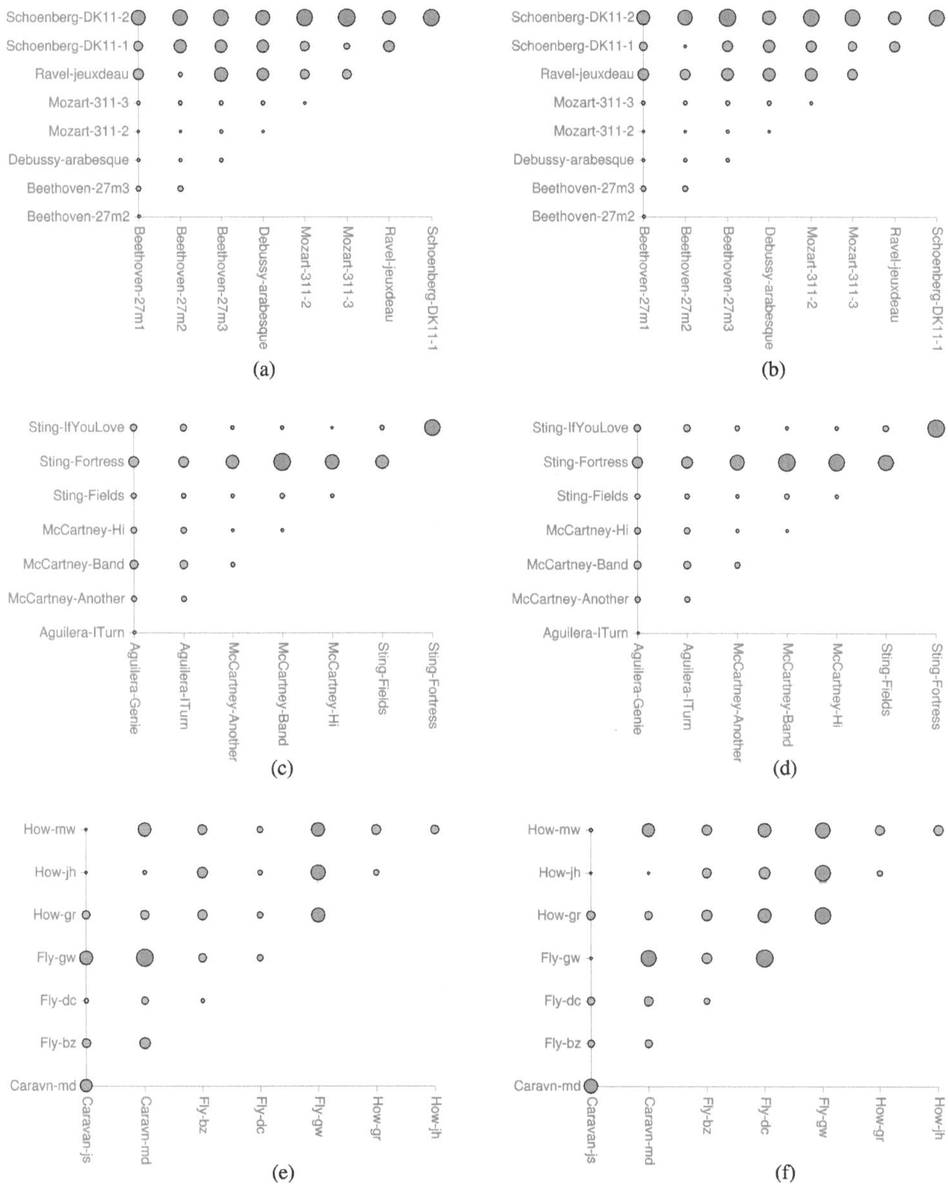

Figure 12. Alignment score of 0-persistence time series for different datasets and variable windowing. The size of the circles associated with each pair of pieces depends on their alignment score. (a) Classic Music (8 bars windowing). (b) Classic Music (4 bars windowing). (c) Pop Music (4 bars windowing). (d) Pop Music (2 bars windowing). (e) Jazz Music (8 bars windowing) and (f) Jazz Music (4 bars windowing).

are obtained by comparing DK11-2 with the compositions by Debussy and Ravel. In general, when compared to the rest of the dataset DK11-1 obtains smaller dissimilarity scores, however, they are sufficient to segregate it from the tonal pieces. The corresponding results, depicted in the distance matrix on the left, do not differ greatly from the one we just discussed. However the tonal traces left in DK11-1 are highlighted by the finer windowing we considered. The same consideration holds for the scores realized by *Jeux d'Eau*. The surprisingly low score generated by its alignment with the second movement of Beethoven's sonata changes by considering a 4-bars windowing. In this case the composition by Ravel is segregated from the others, while

Table 1. Summary of the compositions of the classical music dataset.

Composition	Movements	Author
Sonata n. 27	1, 2, 3	Beethoven
Arabesque		Debussy
Sonata n. 8	1, 2	Mozart
Jeux d'Eau		Ravel
Klavierstück	I, II	Schönberg

Table 2. Jazz dataset. Notation: B.B. arr. stands for big band arrangement, chrom. for chromatic and Man. guit. for manouche guitar.

Label	Ensemble	Style
Caravan-js	4 Gtrs, Org, Kora, Bgtr	B.B. arr., no solo
Caravan-md	Tpt, Pf, Bgtr	Rich solos and tensions
Fly-bz	Bgtr, Vib, Kora, Pf	B.B. arr., chrom. solos
Fly-dc	Flt, Tnr & Bar Sax, F Hn, Org, Gtr	B.B. arr., Man. guit.
Fly-gw	Big Band	B.B. arr.
How-gr	2 Obss, 2 Gtrs, Bgtr	Man. guit.
How-jh	Pf, Bgtr	Chrom. solo
How-mw	Tpt, Gtr, Bgtr, Str, Pf	Embellishments, chrom.

the second movement of the sonata n. 27 obtains a surprisingly low dissimilarity score when aligned with DK11-1. The tonal and pentatonic compositions are highlighted as similar in both representations.

Pop dataset: In both diagrams, the two Aguilera's pieces appear to be well separated from the others. Sting's *Fields of Gold* and *If You Love Somebody Set Them Free* turn out to be similar to the pieces by McCartney. It is not the case for *Fortress Around Your Heart* that recollects high dissimilarity scores when aligned to the other songs of the dataset. It is interesting to note how the two distance matrices are almost invariant with respect to the change of windowing.

Jazz dataset: The classification of jazz standards is a difficult task due to the improvisational nature of this genre. We considered a dataset composed of two versions of *Caravan* and three versions of *Fly Me to the Moon* and *How High the Moon*, respectively. Each interpretation is characterised by different choices in terms of ensemble and arrangements. We summarised these features in Table 2 by denoting a big band arrangement (breaks, horns fills, etc.) as *B.B. arr.*, pointing out the presence of solo parts, their main features, and particular stylistic choices.

The dissimilarity scores computed by considering the global pairwise alignment of the persistence time series associated with these compositions are depicted in the third row of Figure 12. In this example, the information retrieved by the alignment is twofold: on one hand, it stresses the mere melodic and harmonic similarity. On the other hand, it retrieves common stylistic choices. In both distance matrices, the scores associated with the same compositions are reasonably low, highlighting their similarity in the case of *Fly Me to the Moon* and *How High the Moon*. An exception is represented by the two versions of *Caravan*. The presence of rich solos in *Caravan-md* distinguishes it neatly from the other interpretation of the standard. Note how an optimal warping path between these two pieces depicted in Figure 11(b) tries to align them on the themes, skipping the solo parts. Hence, the evolution in time of the persistence diagrams grasps the difference between an organised thematic flow, and a freer improvisational context. Moreover, we notice how the three versions of *Fly Me to the Moon* is well separated from the three versions of *How High the Moon* only utilising a 4-bars windowing. This feature is opposite to the one characterising the analysis of the Pop dataset.

6. Discussion and perspectives

We presented a method to compare time-varying systems by taking advantage of their geometric and topological fingerprint provided by persistent homology.

If the two spaces and the filtering functions are comparable, DTW and an optimal warping path between two kth persistence time series represent the timespans in which the properties determined by the filtering functions, and thereafter fingerprinted as persistence diagrams, are similar. Thus, it gives an encompassing view on the relative geometric behavior of the analyzed systems, by highlighting possible irregular patterns in their geometric evolution.

The effectiveness of this dynamic fingerprint can be tested in many applications, such as animals tracking (Pérez-Escudero et al. 2014; Romero-Ferrero et al. 2019), group behaviour (Munch 2013; Topaz, Ziegelmeier, and Halverson 2015), and classical problems tackled with persistence homology as covering of sensor network (Munch, Shapiro, and Harer 2012) and unfolding of proteins (Cohen-Steiner, Edelsbrunner, and Morozov 2006).

The task of automatic stylistic classification of music is currently tackled in the field of Music Information Retrieval (MIR) (Casey et al. 2008), being an innovative paradigm for the analysis and classification of music compositions. Although the data used to deform the *Tonnetz* in our application are deduced from MIDI transcriptions, thanks to the stability of persistence diagrams with respect to the bottleneck distance, it is possible to generalize the algorithm to audio files, for instance by performing a chroma analysis (Harte and Sandler 2005).

We gave a musical interpretation of the evolution in time of the persistence diagrams associated with a composition and used DTW to provide an alignment of persistence time series. Finally, we analyzed both an optimal warping path and the alignment score of collections of classical (tonal, modal and atonal) compositions, pop songs endowed with different harmonic complexity and a collection of jazz standards played by different ensembles, with different arrangements and solo parts.

In order to interpret the evolution of a system in time as a collection of observations, we defined a windowing, consisting of an even partition of the composition according to its subdivision in bars. We saw how the three datasets we considered respond differently to changes of this windowing. In musical terms, this corresponds to the pace at which musical concepts evolve in the composition. The computation of an optimal time *granularity* necessary to describe the evolution of compositions belonging to different genres or artists could be used as a new music descriptor.

The dynamic *Tonnetz* has full memory of the composition: the heights of its vertices increases monotonically. This feature does not reflect our perception of music, since we cannot remember every note of a whole composition. The definition of a gravity function in opposition to the one generating the deformation of the vertices can be used to endow this space with a type of short-term memory. The same argument can be applied to the study of dynamical consonance-based deformed *Tonnetze*. In addition, we can define a variable gravitational field (or equip the vertices with variable masses) in order to diminish the effect of the gravity in correspondence of vertices representing relevant elements of a musical phrase (for example, its higher, lower, first, ending, and syncopated notes), we refer interested readers to Perricone (2000).

Given the simple structure of the persistence diagrams derived from the *Tonnetz* and the possibility to provide their musical interpretation in this framework, persistence time series could be replaced by continuous vineyards, in order to study the alignment between vineyards by considering the minimal homotopy leading from one to the other.

Acknowledgments

The authors acknowledge the precious comments of Massimo Ferri on an early version of this paper. We thank the reviewers for their careful reading of the manuscript, and their valuable and insightful comments.

Disclosure statement

No potential conflict of interest was reported by the author(s).

ORCID

Mattia G. Bergomi http://orcid.org/0000-0002-3732-3108

References

Bakshi, B., and G. Stephanopoulos. 1994. "Representation of Process Trends-IV. Induction of Real-time Patterns From Operating Data for Diagnosis and Supervisory Control." *Computers & Chemical Engineering* 18 (4): 303–332.

Bergomi, M. G. 2015. *Dynamical and Topological Tools for (Modern) Music Analysis*. Paris: Université Pierre et Marie Curie.

Bergomi, M. G., A. Baratè, and B. Di Fabio. 2016. "Towards a Topological Fingerprint of Music." In *International Workshop on Computational Topology in Image Context*, Marseille, France, 88–100.

Berndt, D. J., and J. Clifford. 1994. "Using Dynamic Time Warping to Find Patterns in Time Series." In *KDD workshop*, 359–370. Vol. 16.

Bigo, L., M. Andreatta, J. L. Giavitto, O. Michel, and A. Spicher. 2013. "Computation and Visualization of Musical Structures in Chord-Based Simplicial Complexes." In *Mathematics and Computation in Music*, 38–51. Berlin: Springer.

Burns, E. M., and W. D. Ward. 1999. "Intervals, Scales, and Tuning." *The Psychology of Music* 2: 215–264.

Casey, M., R. Veltkamp, M. Goto, M. Leman, C. Rhodes, and M. Slaney. 2008. "Content-based Music Information Retrieval: Current Directions and Future Challenges." *Proceedings of the IEEE* 96 (4): 668–696.

Cheveigné, de, A. 2005. "Pitch Perception Models." In *Pitch: Neural Coding and Perception*, edited by C. J. Plack, A. J. Oxenham, A. N. Popper, and R. Fay. Springer, New York, NY.

Cohen-Steiner, D., H. Edelsbrunner, and J. Harer. 2007. "Stability of Persistence Diagrams." *Discrete & Computational Geometry* 37: 103–120.

Cohen-Steiner, D., H. Edelsbrunner, and D. Morozov. 2006. "Vines and Vineyards by Updating Persistence in Linear Time." In *Proceedings of the Twenty-Second Annual Symposium on Computational Geometry*, 119–126. Sedona Arizona, USA. https://dl.acm.org/doi/proceedings/10.1145/1137856

Cohn, R. 1997. "Neo-Riemannian Operations, Parsimonious Trichords, and Their Tonnetz Representations." *Journal of Music Theory* 41 (1): 1–66.

d'Amico, M., P. Frosini, and C. Landi. 2006. "Using Matching Distance in Size Theory: A Survey." *International Journal of Imaging Systems and Technology* 16: 154–161.

d'Amico, M., P. Frosini, and C. Landi. 2010. "Natural Pseudo-distance and Optimal Matching Between Reduced Size Functions." *Acta Applicandae Mathematicae* 109 (2): 527–554.

Douthett, J., and P. Steinbach. 1998. "Parsimonious Graphs: A Study in Parsimony, Contextual Transformations, and Modes of Limited Transposition." *Journal of Music Theory* 42 (2): 241–263.

Edelsbrunner, H., and J. Harer. 2009. *Computational Topology: An Introduction*. American Mathematical Society.

Esling, P., and C. Agon. 2012. "Time-series Data Mining." *ACM Computing Surveys (CSUR)* 45 (1): 12.

Euler, L. 1774. "De Harmoniae Veris Principiis Per Speculum Musicum Repraesentatis." *Novi commentarii Acad. Scient. Petropolitanae* 18, 330–353, repr. in Leonhardi Euleri, Opera Omnia, Series III, T.1, Teubner, Liepzig, Berlin 1926.

Ferri, M. 2017. "Persistent Topology for Natural Data Analysis – A Survey." In *Towards Integrative Machine Learning and Knowledge Extraction*, 117–133. Cham, Switzerland: Springer.

Govc, D. 2013. "On the Definition of Homological Critical Value." arXiv:1301.6817.

Harte, C., and M. Sandler. 2005. "Automatic Chord Identifcation Using a Quantised Chromagram." In *Audio Engineering Society Convention*. Audio Engineering Society, New York, Vol. 118.

Hatcher, A. 2002. *Algebraic Topology*. Cambridge: Cambridge University Press.

Keogh, E., S. Chu, D. Hart, and M. Pazzani. 2004. "Segmenting Time Series: A Survey and Novel Approach." *Data Mining in Time Series Databases* 57: 1–22.

Keogh, E., and S. Kasetty. 2003. "On the Need for Time Series Data Mining Benchmarks: A Survey and Empirical Demonstration." *Data Mining and Knowledge Discovery* 7 (4): 349–371.

Liao, T. W. 2005. "Clustering of Time Series Data – a Survey." *Pattern Recognition* 38 (11): 1857–1874.

Lines, J., and A. Bagnall. 2015. "Time Series Classification with Ensembles of Elastic Distance Measures." *Data Mining and Knowledge Discovery* 29 (3): 565–592.

Munch, E. 2013. *Applications of Persistent Homology to Time Varying Systems*. Durham, North Carolina: Duke University.

Munch, E., M. Shapiro, and J. Harer. 2012. "Failure Filtrations for Fenced Sensor Networks." *The International Journal of Robotics Research* 31 (9): 1044–1056.

Munkres, J. R. 1984. *Elements of Algebraic Topology*. Vol. 2. Cambridge Massachusetts: Addison-Wesley Reading.

Perricone, J. 2000. *Melody in Songwriting: Tools and Techniques for Writing Hit Songs*. Boston: Hal Leonard Corporation.

Pérez-Escudero, A., J. Vicente-Page, R. C. Hinz, S. Arganda, and G. G. Polaviejade. 2014. "idTracker: Tracking Individuals in a Group by Automatic Identification of Unmarked Animals." *Nature Methods* 11 (7): 743–748.

Poincaré, H. 1895. *Analysis Situs*. Paris, France: Gauthier-Villars.

Romero-Ferrero, F., M. G. Bergomi, R. C. Hinz, F. J. Heras, and G. G. Polaviejade. 2019. "idtracker. ai: Tracking All Individuals in Small Or Large Collectives of Unmarked Animals." *Nature Methods* 16 (2): 179.

Senin, P. 2008. *Dynamic Time Warping Algorithm Review*, 40. Information and Computer Science Department University of Hawaii at Manoa Honolulu, USA 855.

Topaz, C. M., L. Ziegelmeier, and T. Halverson. 2015. "Topological Data Analysis of Biological Aggregation Models." *PloS One* 10 (5): e0126383.

Turner, K., Y. Mileyko, S. Mukherjee, and J. Harer. 2014. "Fréchet Means for Distributions of Persistence Diagrams." *Discrete & Computational Geometry* 52 (1): 44–70.

Žabka, M. 2009. "Generalized Tonnetz and Well-Formed GTS: A Scale Theory Inspired by the Neo-Riemannians". In *International Conference on Mathematics and Computation in Music*, edited by C.A.C.C.H. Chew Elaine, 286–298. Berlin: Springer.

Index

Note: **Bold** page numbers refer to tables; *italic* page numbers refer to figures and page numbers followed by "n" denote footnotes.